STOP GARBAGE

The truth about recycling

Alex Pascual

Edition: February 2019

© Alex Pascual
www.stopgarbage.com

All rights reserved. No part of this publication may be reproduced, stored in a retrieval system, transmitted in any form or by any means, electronic, mechanical, photocopying, recording or otherwise, without the prior written permission of the author.

Cover Designer: Ignacio García Bermúdez

Text revision: Amy Sue Bennett

ISBN: 978-17-951-8335-2

Printed by Independently published, an Amazon.com Company

*For my parents, Montse and Juan,
for my friends and for those who
believe that we can leave a better
world to the generations to come...
Among which are my dear nephews
and nieces: Joan, Carla, Rita and Ramón*

CONTENTS

1. Presentation 9
    1.1. Why did I write this book? 9
    1.2. Why am I qualified to write this book 12
2. How much garbage do we produce? 15
    2.1. What is our garbage like? 17
3. Where does the garbage go and how much do we recycle? 19
4. The waste problem 27
5. What should we to do with waste? 33
    5.1. Landfill 35
    5.2. Ecopark: a step before landfill or incineration 41
    5.3. Incineration 45
6. Recycling is the solution 57
7. How important is recycling? 65
    7.1. Glass: a successful case study 65
    7.2. Paper comes from the trees 71
    7.3. The importance of recycling plastic and metal (packaging) 79
    7.3.1. The kidnapping of the yellow container 91
    7.3.2. The biggest plastic dump in the world: The ocean 97
    7.4. Biowaste: A key point 101
    7.4.1. Food waste 107
    7.5. The benefits of waste recycling: A summary 111
8. The best waste is no waste 115
9. Global warming and waste 121
10. Economy and waste 127
    10.1. GDP and waste 128
    10.2. Recycling is cheaper than not recycling 133
    10.3. Recycling creates employment: less garbage, more employment 139
    10.4. Waste as resources: moving towards a circular economy 145
Epilogue: Recycling in brief 151

1.
PRESENTATION

1.1. Why did I write this book?

I've always have been into environmental awareness. Everyone has their own concerns, I guest.

On a professional level, I've had different works related to the environment (waste and water) and, more specifically, with garbage and recycling.

Also, my current position as a municipal technician has enabled me to observe and check exactly what is recycled in each container or bin, the amount of waste and way this industrial sector, which is very professionalized indeed, behaves.

My professional vision, far from making me a technical specialist in waste treatment, gives me an overview of the global waste management process; how people or households recycle, how we use recycling bins, where or in which treatment facilities the garbage ends up, and in what quantities. This has provided me with knowledge that enables me to explain the current waste recycling options available to us (especially in Europe).

At work, I have to tell people to recycle because it's good for the environment and I reel off the typical and topical speech... That's when I'm overcome with doubts: *Is recycling really critical? Why? What impact does waste recycling have on the environment? What implications does it have on water, on land, on the atmosphere and its resources? Does it affect the environment? To what extent? And what about incineration? Is it harmful? How about landfills? Does recycling our waste generate more jobs? What impact does the recycling industry have on the economy?* The truth is that communication available in this field does not usually quantify and explain the recycling benefits successfully. That's why I started to research the reasons and facts or *the truth about recycling*.

Is recycling really important?

From a personal and social point of view, I see friends and family around me, as well as many people from different backgrounds (educated, or not), different professions (designers, engineers, artists, administrators, workers, among others), of different ages—especially young people born in our country in the heart of democracy— educated in the values of ecology and respecting the environment and, quite simply, they do NOT recycle. They're not interested in recycling. They say that recycling is dirty. Garbage "stinks". Some justify themselves by saying: "Why should I recycle when the government mixes all the recycling bins or containers up afterwards anyway?"–an urban legend and, by the way, a big fat lie. Other people say: "I don't have room for so many different recycling bins." In my opinion, they're all just excuses, different ways of appeasing their conscience.

So, because of these doubts and, above all, the need to explain to those who do not recycle why it's essential to do so, I've written this book.

I will try to explain the recycling "world" in a pleasant way but, at the same time, with sufficient information so as to understand *what, how much and why* it is crucial to recycle our waste, based on verified data and above all with sufficient criteria for knowing which sources are truthful, official and help to easily explain the information.

The objective is not only "to care for the environment" but also to care for the only planet that we have (there is no plan B), and to show that recycling affects us directly, along with our quality of life, future generations and, ultimately, that *it makes us happier* with ourselves and with our actions.

The book is written in a direct style, as if the reader and I were talking face-to-face with me explaining what I know about waste recycling.

I hope you like it and that it awakens your curiosity, at least. Enjoy your read!

1.2. Why am I qualified to write this book?

Let me briefly go over my background to explain why I am able to write this book.

I studied a bachelor's degree in Industrial Engineering at the Polytechnic University of Catalonia (UPC) and, given my environmental awareness, I took various subjects related with the environment and I also did a summer internship on a garbage truck for the Sant Cugat del Vallés City Council carrying out a quality control audit.

I believe that this summer job opened doors for me and, as a result, I was hired by the environmental engineering consulting services where I worked for almost five years and was involved in various projects for the Barcelona City Council, supporting the strategic municipal waste management plan and the new public tender for waste collection and street cleaning services. An exciting world, and despite dealing with garbage, a very professional sector!

However, I have learned the most about *real* waste management and the importance of recycling during the more than nine years that I have been working at El Prat del Llobregat City Council, managing public services and different cleaning and waste contracts. This job has given me knowledge in the field and has made me aware of the importance of the work carried out by the public administration along with the excellent work that many officials do for all of us. That's why I want to share it with you.

My commitment is to speak clearly and with clarity about recycling issues, without taking sides myself but reaching conclusions, with reliable and contrastable data, backed up by different official, municipal, autonomous, state, European and international organizations.

I have tried to be as non-technical as possible, to present three critical points in each chapter and to shorten some aspects to make for fluent reading. I hope that my colleagues in the sector don't hold it against me.

To see my professional profile on LinkedIn, visit:

es.linkedin.com/in/alexpascualc/

2.
HOW MUCH GARBAGE DO WE PRODUCE?

When we talk about garbage, we're usually referring to waste that we produce directly and which is formally known as "household" or municipal waste. But we mustn't forget that we generate other waste indirectly, be it industrial waste —from consumer goods —during manufacturing, or construction waste from real estate. Therefore, municipal waste only accounts for about one third of the total waste generated, just over another third corresponds to industrial waste (including sewage sludge) and less than a third is construction or demolition debris waste[1]:

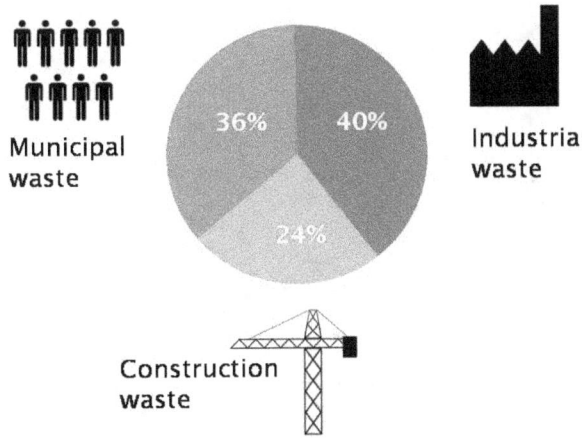

---

1. Data for Catalonia 2012. General prevention and waste management of Catalonia 2013-2020

Municipal waste only makes up 1/3 of the total waste generated

In Catalonia,[2] in 2012 the waste generated was over 10 million tons including municipal, industrial and construction debris. In Spain,[3] in 2012, almost 120 million tones were gererated and in the same year in Europe[4](EU27) more than 2,500 million tons. That's a lot of waste! Especially if you consider that 1 ton is equal to 1,000 kg.

---

2. *Ibidem* (same as previous quote)
3. EUROSTAT. Official statistics. European Union
4. *Ibidem*

2.1. What is our garbage like?

The waste that ends up in street containers or bins comes mainly from households and businesses (which contribute about 1/3 of municipal waste). If you review the type of waste found in the general waste or trash[5] container (usually the gray one) and add the waste collected from the recycling bins, the result is the following standard garbage bag:[6]

30% other
8% glass
12% plastic and metals
12% paper and cardboard
38% biowaste

38% of the waste the we generate is organic

---

5. Technically known as general waste or household waste (UK) or trash (USA), the waste fraction that is left over once the other recycling fractions have been removed, we can find several different types: diapers, pads, cleaning waste, broken pottery, cigarette butts and ashes, among others
6. My compilation from different studies: "Pesa la brossa" 2014. Study by the Polytechnic University of Catalonia and General Program of Prevention and Waste Management of Catalonia 2013-2020. According to the Agència de Residus de Catalunya 2014, the data are organic 37%, paper and cardboard 12%, glass 8%, plastics and metals 12%. "La gestió dels residus i el seu impacte en el canvi climàtic." Statistics 2014

Most of the garbage that we generate is organic matter and comes from household food waste and green waste. It represents 38% of the total waste, although this varies depending on the region or the country. Subsequently, in developing countries, this percentage is higher, while in more developed countries it is lower.

Glass, plastic, metal, cartons, paper and cardboard packaging represent 32%. The remaining 30% corresponds to various materials – such as clothing, furniture, demolition debris– that could be recycled via other channels. It is estimated that 84% of household waste is recyclable.

3.
WHERE DOES GARBAGE GO AND HOW MUCH DO WE RECYCLE?

All of the garbage that we generate can be distributed between different containers or bin types according to the garbage collection model of each city. However, once collected, waste can generally be disposed of in 3 different ways: recycling (including composting), landfill or incineration. I do not include reused objects or waste since they obviously don't end up in a treatment facility but rather in the hands of another person who uses them or gives them a new use.

Garbage ends up in recycling facilities, landfills or incinerators

Of the total amount of municipal waste treated in Catalonia[7] in 2012, 46% went to landfill, 18% was incinerated and 36% was sent to recycling facilities (including composting facilities). In Spain,[8] in 2012, 61% went to landfill, 9% was incinerated and 30% to recycling facilities (recycling and composting). In Europe[9](EU27), in 2012, 33% ended up in landfill, 25% was incinerated and 42% was sent to recycling facilities (recycling and composting). In the cities of the world,[10] it is estimated that 56% of the waste goes to landfill, 17% for incineration and 27% is recycled (recycled and composted).

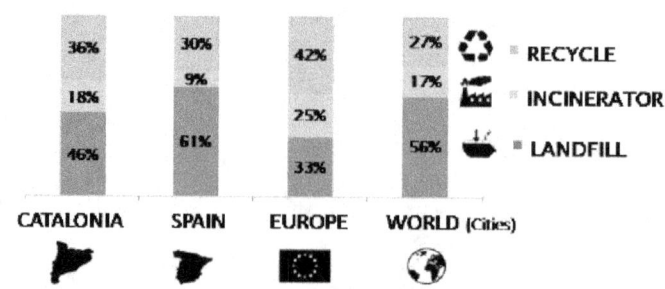

---

7. Based on data extracted from the General Program for the Prevention and Management of Waste of Catalonia 2013-20. The data does not account for 100% of the waste due to a reduction in weight
8. EUROSTAT. Official statistics. European Union
9. *Ibidem*
10. "What a Waste," World Bank report, 2012

In the case of Catalonia, it's worth specifying that selective recycling collection is at 39% and that, due to the wrong waste being deposited in the recycling bins (inappropriate items), this percentage is reduced to 32% (net recycling rate). Afterwards, this rate is bumped up by the waste recovered from Ecoparks. Therefore, the final recycling figure would be around 36% (data compiled from other official data).

In light of these recycling quotas, we should be aware of the **low level of waste recycling** being carried out both by individuals and businesses. If selective waste collection is at 39% in Catalonia, which means that many people are participating in selective collection and recycling, but it also means that there are still those who do not recycle at all. The figures speak for themselves: we haven't even achieved 50% when it comes to recycling.

More than half of our waste ends up in landfills and incinerators

In the case of Spain,[11] only 15% of waste is collected separately. In other words, 85% of the waste is collected mixed in the same container (general waste or trash). The recycling rate increases to 30% thanks to compost recovery in the Ecopark from the general waste fraction. However, the compost produced from the general waste container is of worse quality than that generated in a composting facility with selective collection from the organic container.

---
11. Annual report 2013 Ministerio de Agricultura, Alimentación y Medio Ambiente, Spanish Goverment

As for the European (EU27) data,[12] this is shown in the following table, for European countries ordered by the highest recycling rate for the year 2012. I don't want to overwhelm you with data and tables, but I do think that this is worth taking a look at:

COUNTRY	LANDFILL	INCINERATION	RECYCLING
GERMANY	0%	35%	65%
AUSTRIA	5%	36%	59%
BELGIUM	1%	42%	57%
NETHERLANDS	2%	49%	49%
SWEDEN	1%	51%	48%
LUXEMBOURG	17%	35%	48%
SLOVENIA	51%	2%	47%
DENMARK	2%	54%	44%
UNITED KINGDOM	38%	19%	43%
EU (27 countries)	33%	25%	42%
IRELAND	42%	18%	40%
ESTONIA	44%	16%	40%
ITALY	41%	19%	40%
FRANCE	29%	34%	37%
FINLAND	33%	34%	33%
SPAIN	61%	9%	30%
BULGARIA	73%	0%	27%
PORTUGAL	55%	19%	26%
HUNGARY	66%	9%	25%
POLAND	75%	0%	25%
LITHUANIA	76%	0%	24%
CZECH REP.	57%	20%	23%
CYPRUS	78%	0%	22%
GREECE	81%	0%	19%
LATVIA	84%	0%	16%
SLOVAKIA	76%	10%	14%
MALTA	87%	1%	13%
ROMANIA	97%	0%	3%

---

12. EUROSTAT. Official statistics. European Union

Where does the garbage go and how much do we recycle?

In the most advanced European countries, 2% of waste ends up in landfill and 43% is incinerated

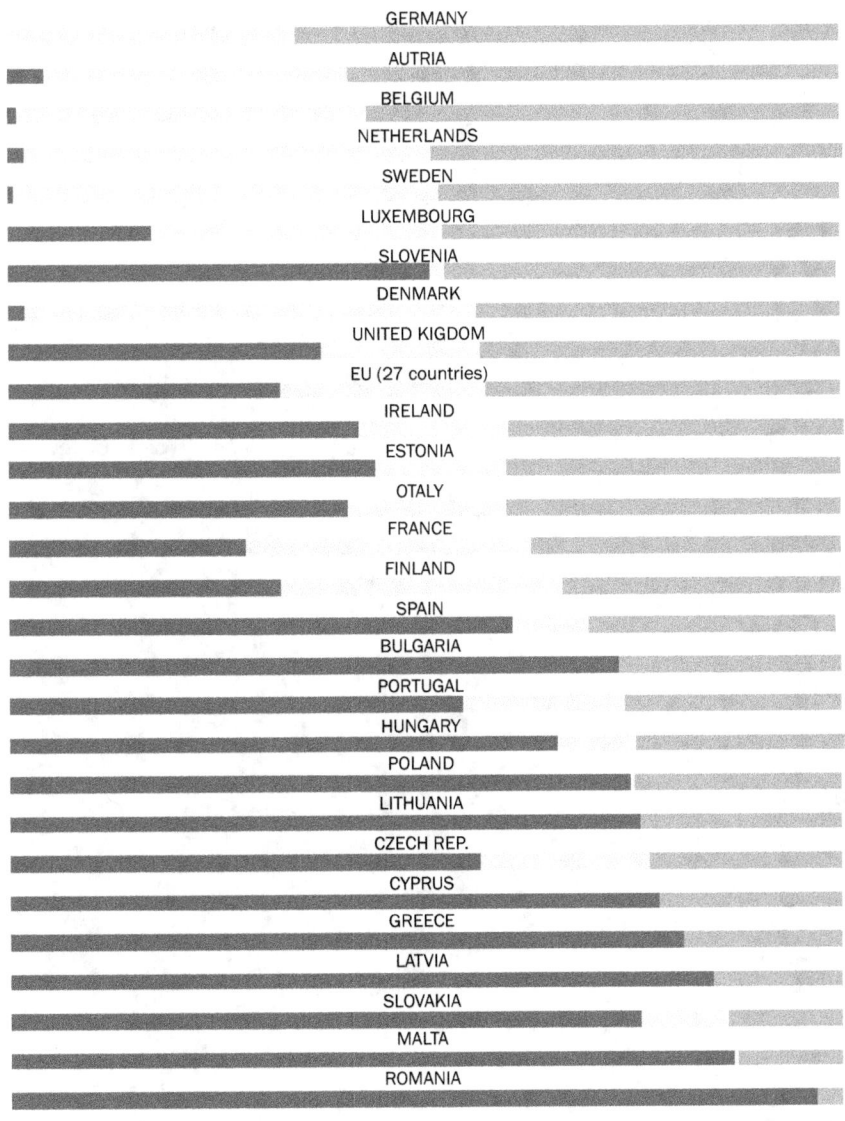

LANDFILL   INCINERATION   RECYCLING

23

From this table, it is clear that the European countries with more experience in waste treatment – such as Holland, Belgium, Germany, Austria or Sweden –send only, on average, 2% of waste generated to landfill but, on the other hand, they have higher incineration rates –43% on average– and recycle 56% of their waste.

Want to know more?

♟ I recommend visitingt the Waste Atlas website where there is a lot of information about municipal waste in many countries and cities around the world (ENG)

http://www.atlas.d-waste.com/

♟ Statistical data EUROSTAT of the European Union (ENG)

http://ec.europa.eu/eurostat/statistics-explained/index.php/Municipal_waste_statistics

4.
THE WASTE PROBLEM

This book focuses on domestic or municipal solid waste (MSW), which is the waste that we generate directly as a result of our behavior, individually or as a family, and regarding which we have more capacity, freedom, and possibility to influence in terms of management or recycling.

Among the different existing definitions of waste, the most thorough could be the following: "Any substance or object which the owner discards, has the intention of discarding or is obliged to discard. The reason for wanting to discard said substance or object is usually associated with the fact that its owner considers it not valuable enough to retain."

Stop garbage

> Waste is any substance or object that its owner discards, intends to discard or is obliged to discard

Waste generation has increased in massive proportions over the last 100 years, but this has not always been the case. The causes of this increment could be attributed, on the one hand, to the development of society, to modernization or the consumption system. And, on the other hand, this increase is reinforced by the exponential growth of the population.

The following table shows the evolution of garbage generation in the USA[13] over a wide range of years, from 1960 to 2011:

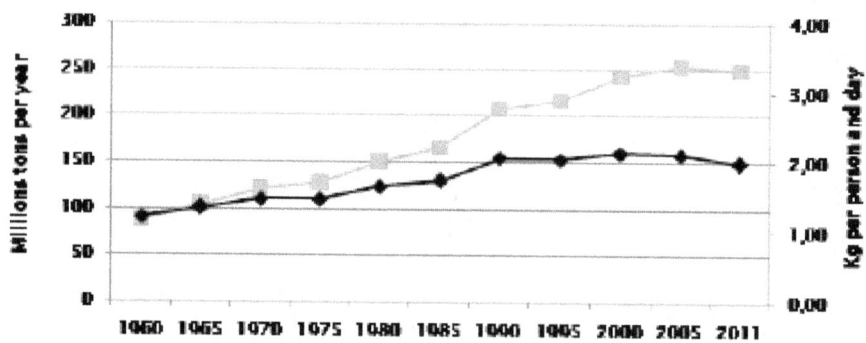

> In the USA, municipal waste has tripled in the last 50 years

---

13. United States Environmental Protection Agency. Municipal Solid Waste Generation, Recycling, and Disposal in the United States: Facts and Figures for 2012

Here, we can see that the municipal waste has almost tripled in only 51 years. The same proportion could be applied to Europe (EU27) if there were data before the EU. From 1995 to 2003, the municipal waste in the EU25 increased[14] by 19% (the same percentage as the economic activity or GDP). In Spain, waste generation increased[15] by 55% between 1990 and 2003.

Let's take a look at the amount of municipal waste that we produce or generate in different areas:

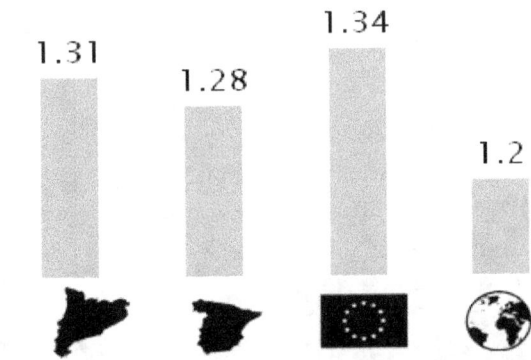

	CATALONIA	SPAIN	EUROPE	WORLD*
POPULATION (Millions)	7.5	46	434	2,968
WASTE GENERATED (Millions tons/year)	3.58	21.17	212.22	1,300.00
WASTE X PERSON & YEAR (kg/persona and day)	1.31	1.28	1.34	1.20

*WORLD: Only cities

---

14. Libro Verde *de la sostenibilidad urbana y local en la era de la información*, Ministerio de agricultura, alimentación y medio ambiente, 2012
15. *Ibidem*

Regarding the **total amount** of waste, in Catalonia,[16] in 2012, almost 3.6 million tons of municipal waste was generated. The same year in Spain,[17] over 21 million tons were generated in Europe[18] (EU27), over 212 million tons and in the cities of the world,[19] a total of 1,300 million tons was estimated in 2012.

Concerning **quantity per inhabitant**, in Catalonia, in 2012, over 1.31 kilos of municipal waste was generated per inhabitant per day. In Spain, almost 1.28 kilos per inhabitant per day were produced. In Europe (EU27), around 1.34 kilos and the cities around the world, a generation of 1.20 kilos per inhabitant per day is estimated.

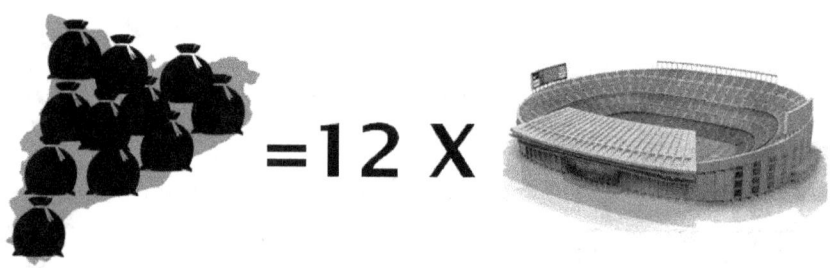

With almost 3.6 tons of municipal waste generated per year in Catalonia, Camp Nou, the F.C. Barcelona stadium, could be filled 12 times (70 stadiums with all the waste generated in Spain).

---

16. Agència de residus de Catalunya, Statistics 2012
17. Annual report 2013 Ministerio de Agricultura, Alimentación y Medio Ambiente, Spanish Goverment and EUROSTAT
18. EUROSTAT 2012. Official statistics. European Union
19. "What a Waste," Informe del World Bank (2012)

Causes[20] for the increase in waste are varied but the following can be highlighted:

- The inefficiency of production systems (each ton of waste used or consumed can generate 20 tons of raw material waste in the extraction phase and 5 tons of waste in its production)
- The reduction of the useful life of products or the increase in planned obsolescence that occurs for many reasons, such as quality, fashion or technology or the appearance of new functionalities
- Compulsive product buying
- Failure to incorporate environmental and social costs into the cost of goods or products
- The orientation of marketing strategies towards the increase, in both quantity and diversity, of product packaging

To add to the causes for this increase in garbage per person, the number of inhabitants in the world is rising, and subsequently, the total amount of waste produced also rises.

---

20. Libro Verde *de la sostenibilidad urbana y local en la era de la información,* Ministerio de Agricultura, Alimentación y Medio Ambiente, 2012

Causes for increase in waste: marketing, increase in population, product obsolescence

There is evidence[21] of a link between economic progress and the waste generation (*see chapter 10*).

Whatever the cause, there's no doubt that the increase in waste has not only been extraordinary, but continues to increase in many countries - developed or in the OECD[22]- which generate 44% of the world's waste. What will happen when developing countries such as China (which is already the second highest country in waste production, after the USA) —or India—achieve the same waste generation quotas per inhabitant? It is estimated[23] that, in 2025, global waste generation in cities around the world will rise to levels of around 4,300 million tons per year, which would see, in just 13 years, the current waste figures triple.

Waste is an environmental, social and economic problem

The —ecological, social, and economic— problems that waste causes will undoubtedly worsen in the years to come if we continue to fail to recycle (reduce, reuse, compost). With this in mind, we should ask ourselves the next question…

---

21. See chapter 10, "Economy and waste."
22. The organization for economic co-operation and development (OECD): Countries (resume): European Union, USA, Canada, México, Chile, South Korea, Israel, Australia, New Zeeland, and Japan. Percentage from "What a Waste," World Bank, 2012
23. "What a Waste," World Bank report, 2012

5.
WHAT SHOULD WE DO WITH WASTE?

Every year, worldwide, cities around the world are generating around 1,300 million tons of waste (about 3.6 million tons per day, bearing in mind that one ton is 1,000 kg). Also, this amount of garbage increases over the years. What should we do, then, with such a high amount of waste? Do we bury it? Do we burn it? Or do we recycle it?

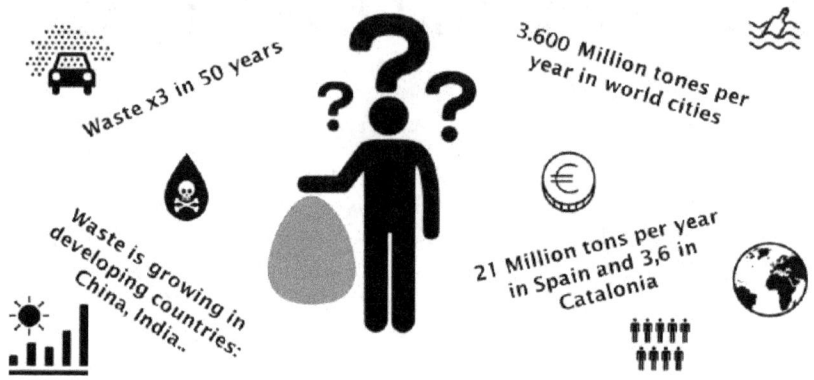

What should we do with our garbage

The history of humanity shows that the destination or the main solutions to this question have been landfills and, over the last 150 years, also incineration. Therefore, most of the waste that we generate ends up in these treatment facilities. However, what does each one consist of? What impacts do they have on the environment and people? Let's look at an in-depth review of each one.

Want to know more?

1 The world's trash crisis, and why many Americans are oblivious, *Los Angeles Times* (ENG)

http://www.latimes.com/world/global-development/la-fg-global-trash-20160422-20160421-snap-htmlstory.html

5.1. Landfill

From the historical point of view, landfills are the leading and oldest way to treat waste and probably constitute the first waste management system. In Banyoles (Girona) there is a Neolithic site (La Draga) which is over 7,300 years old with pits that were used as landfills.

The simple accumulation of waste discharges, traditionally known as dumps, has evolved into a landfill-controlled deposit, where we try to minimize the impact that garbage accumulation has on our environment. Landfills – or sanitary landfills– are waste disposal facilities which are either superficial or underground. They are waterproofed, the gas emitted is (partially) captured and waters are collected.

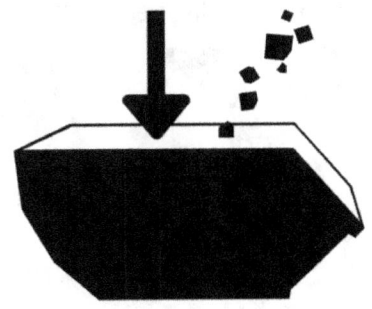

Landfills dispose of waste superficially or underground

There are three different classes of landfill, according to the waste type deposited in them: the nonhazardous landfill, hazardous landfill and those for inert[24] waste (demolition or construction debris). I'm going to center my review in the nonhazardous landfill, since, as I have said, this is where most of the municipal waste from households and businesses ends up.

Of the total waste generated in 2012 in Catalonia,[25] the 46% ended up in landfills, in Spain,[26] this figure was 61%, in Europe[27] (EU 27), only 33% and in the main cities around the world,[28] 56%.

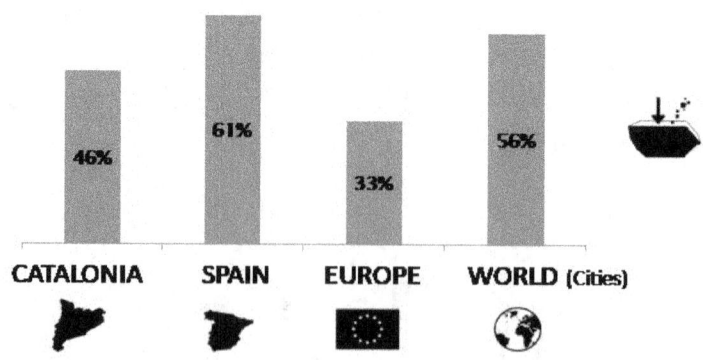

---

24. Residues that do not undergo any physical-chemical or biological transformation
25. Data extracted from Prevention and waste management general program, Catalonia 2013-2020. Data does not account for 100% due to the weight reduction
26. EUROSTAT. Official statistics. European Union
27. *Ibidem*
28. "What a Waste," World Bank report, 2012

There is an extensive landfill network to treat such quantities of waste: 31 located in Catalonia[29] and a total of 134 in Spain.[30]

However, managing waste by directly sending it to landfill is **the least recommended option due to the impacts this can cause.**[31] The main problems with municipal waste landfills are focused on:

- The decomposition and fermentation of the organic matter present in the waste generates biogas, mostly a mixture of methane gas (CH_4) and carbon dioxide (CO_2), which are responsible for global warming
- Methane can accumulate in the landfill and could cause an explosion
- Soil and water contamination in the region caused by the water from the landfill draining various contaminating materials (heavy materials such as lead, mercury, among others, that may be present in the waste)
- Land occupation and impact on the landscape

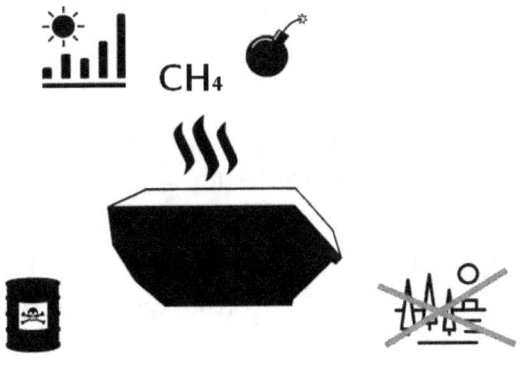

29.. PINFRECAT: Territorial plan for municipal waste management infrastructures in Catalonia 2013-2020
30. Annual report 2013 Ministerio de Agricultura, Alimentación y Medio Ambiente, Spanish Government
31. "Being wise with waste: the EU's approach to waste management," European Commission, 2010

Landfill is the least recommended option for treating waste due to its impacts

With the aim of reducing the environmental impact on the planet, modern landfills are waterproofed to prevent the polluting products from migrating or transferring to the environment. Likewise, in landfills, liquids present in the waste that carry toxic or polluting materials (leachates) are channeled. Any biogas produced is also collected, which is used, for example, to generate electricity.

Despite the preventive measures applied, landfills emit such a high quantity of methane gas–CH_4– that they constitute the primary source of greenhouse gas emissions in waste treatment. I have always heard that there's a gas capture system in these facilities, but what I didn't know is that said catchment[32] is only estimated to be effective for 20%[33] (19%[34]) of the methane that they emit.

Landfill is the main cause of greenhouse effect emissions (methane CH_4) in waste management

In this sense, and to avoid greenhouse gas emissions, the European Union established a Directive[35] for 2016 that obliges all member countries to reduce the biodegradable waste or biowaste (mostly organic matter) present in landfills to 35% of the organic waste generated in 1995.

---

32. Prevention and waste management general program, Catalonia 2013-2020
33. Data extracted from: "Emission evolution in Catalonia" Catalan office for climate change, 2014
34. "A Changing Climate for Energy from waste, for friends of earth," Dominic Hogg report for the well-known consultancy Eumonia
35. Directive 1999/31/CE, article 5

Also, another European Directive[36] forbids waste from being sent directly to landfills if it has not received previous[37] treatment in Ecoparks (more information in the next chapter).

These two regulations are good examples, among many others, of the benefits that come with belonging to the European Union, benefits that we sometimes take for granted.

I wouldn't like to end this chapter without saying that, thanks to advances in technology and legislation, in Spain, in Europe and the West in general, dumps are landfills or controlled waste deposits since they are waterproofed and have a system for capturing biogas, among other measures. However, in Spain, it is still estimated[38] that 4% of waste is not treated in any way and is dumped in an uncontrolled and illegal manner in fields and mountains. Some sad examples are the Abanilla landfill or the one in Campoamor (which is closed now, but it is still contaminated). And let's not forget that the European Commission denounced the Spanish[39] State for having 61 illegal dumps.

Worldwide, it is estimated[40] that 10% of urban waste ends up in uncontrolled dumps. You can see the environmental impact of these uncontrolled discharges in the shocking documentary **TRASHED**, where a dump in Syria directly contaminates the Mediterranean Sea.

---

36. Directive 98/2008/CE) Transpose in the law Ley de residuos y suelos contaminantes, Spanish Government
37. With some exceptions
38. Libro Verde *de la sostenibilidad urbana y local en la era de la información,* Ministerio de Agricultura, Alimentación y Medio Ambiente, 2012
39. "Brussels denounces Spain for not eliminating 61 illegal dumps," *El País,* 16 July 2015
40. "What a Waste," World Bank, 2012

Want to know more?

▶ Award-winning documentary TRASHED, starring Jeremy Irons. Very instructive, easy to understand, with good images, good content and good technical details (ENG)

http://www.trashedfilm.com/

▶ *Basureros para rato, RTVE. Program El Escarabajo Verde,* a documentary about illegal dumps (*CAST*)

http://www.rtve.es/television/20150422/basureros-para-rato/1134406.shtml

♟ Europe denounces Spain for having 61 illegal landfills (ENG)

https://www.endswasteandbioenergy.com/article/1427805/eu-court-declares-61-spanish-landfill-sites-illegal

♟ Landfill clousure in Menorca due to leachate leakage (CAT)

https://directa.cat/actualitat/clausurat-lunic-abocador-de-menorca-filtracions-de-liquids-toxics-mila

5.2. Ecopark: a step before landfill or incineration

Ecoparks are modern waste treatment facilities that allow some recyclable materials to be recovered and reduce the amount of waste destined for landfills or incineration.

Ecopark 1 in Barcelona, which opened in 2003 along with Pinto[41] Ecopark (Madrid), was one of the first integrated waste treatment centers to open in Spain.

In an Ecopark, the mixed or general waste is subjected to what is known as Mechanical and Biological Treatment (TMB). "Mechanical," because the waste is separated and classified by mostly mechanical processes that allow recyclable materials, such as metals, plastics, glass... to be recovered. "Biological," because the waste left over after the mechanical stage, with a high presence of organic matter, is treated by biological processes.

---

41. Xavier Elias Castells. *Métodos de valoración y tratamiento de los residuos municipales,* Díaz Santos, 2012

We have seen the impact on the environment caused by the presence of organic matter – or biodegradable waste – in landfills, and the problem of leachate these days. Today, this impact is inevitable. With the current recycling rates, for a while, there will still be mixed waste that needs to be treated. Treating the mixed waste beforehand in **an Ecopark is the best option to avoid the greenhouse effect produced by the waste in landfills,**[42] since it allows as much as possible of the recyclable waste present in the mixed fraction (biodegradable materials, metals, plastics, paper, glass, etc.) to be recovered and, this way, only the material that cannot be recovered (stabilized material) is sent to landfill.

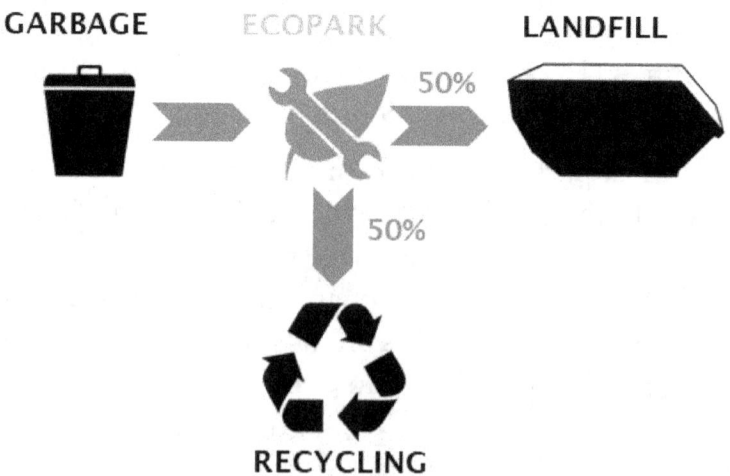

Waste treatment in an Ecopark reduces greenhouse emissions caused by landfill

---

42. "A Changing Climate for Energy from waste, for friends of earth," Dominic Hogg report for the well-known consultancy Eumonia

What should we do with waste?

In 2012, in Catalonia, 50% of the waste from the mixed fraction or the gray container was pre-treated (MBT or mechanic and biological treatment in an Ecopark), 11% was incinerated, and 39% was sent directly to landfill.[43] The Generalitat of Catalonia is working with the objective that, in 2020, the entire mixed fraction will receive pre-treatment before being sent to landfill.[44]

> Ecoparks allows to amount of waste sent it to landfill or incineration to be reduced by half

Ecoparks or MBT facilities allow the mixed fraction, that would be sent straight to landfill, to be reduced. In Catalonia, this is currently 52%. More than a half! This is achieved by recovering materials (metal, plastics, cartons, etc.), obtaining a stabilized material, resulting in waste reduction, among other things.

Ecoparks are waste facilities that, despite not being very well known, constitute an essential link in the entire waste management chain, contributing significantly to environmental conservation. If you have the possibility to visit this kind of facility in your municipality, don't hesitate to go. It's well worth a visit.

---

43. Programa General de Prevención y Gestión de residuos de Cataluña 2013-2020. Dato de vertedero corregido con las estadísticas de la Agència de Residus de Catalunya 2012
44. *Ibidem*

Want to know more?

❶ Visiting program for Ecoparks in Barcelona (ENG)

http://www.amb.cat/en/web/medi-ambient/agenda

▶ Ecopark 4 in Hostalets de Pierola (CAST)

https://www.youtube.com/watch?v=W8gOx01RNIo

❶ What is an Ecopark? (ENG)

http://residus.gencat.cat/en/ambits_dactuacio/valoritzacio_reciclatge/instal_lacions_de_gestio/ecoparc/index.html

5.3. Incineration

There's no doubt that the burning of waste has been common throughout history. However, it wasn't until 1874, in Leeds (England), that the first urban waste incinerator was built, due to a cholera epidemic. Alfred Fryer was the inventor who designed the incinerator for purifying organic matter and, curiously, he called the invention: Destructor.

Waste incineration is a well-known and popular alternative for treating the waste that we generate by means of destruction, waste-to-energy or energy recovery.

In an incineration facility, controlled combustion occurs at high temperatures (over 850°C) to treat the mixed or general fraction as well as rejections from other treatment facilities, such as those from packaging facilities or Ecoparks (material that can't be recycled). All of the material that enters the incineration facility is transformed into ashes, slag, and gases. The energy that is produced during combustion or waste burning can be converted into electricity (heating water and with a turbine) or used for air conditioning (heating and cooling).

Waste is burned in a controlled manner in an incinerator to produce energy

Not all waste is burned at the incineration facilities as some of the materials do not reach their melting point or residual products are created, among which:

- Slag or bottom ashes. This is the material that remains unburned in the oven after combustion, such as ceramics, soil, glass, metal objects, among others. This represents 20-25%, in weight, of the incinerated waste. Slag is often reused in other sectors, and non-metallic slag is, when possible, used as filler material. It is qualified as non-hazardous waste.

- Fly ashes (volatile materials). They represent around 2-6%, in weight, of the incinerated waste. This waste is more dangerous and polluting than the previous types and is collected separately to be taken to hazardous landfill. It is classified as hazardous waste.

Incineration facilities belong to the energy recovery facilities or waste-to-energy group, although there are other processes through which energy is obtained, such as:

- Incinerators
- Anaerobic digestion in biological mechanical treatment facilities or MBT (Ecoparks)
- Fuels derived from waste: material which is prepared and selected and subsequently replaces a fossil fuel
- Landfills or controlled deposits: biogas is obtained from the fermentation of organic matter

It is worth mentioning the latest generation of waste-to-energy, such as the transformation of waste into fuel or RDF (Refuse-derived fuel), which can be used, for example, in cement facilities; and the production of ethanol or diesel from waste.

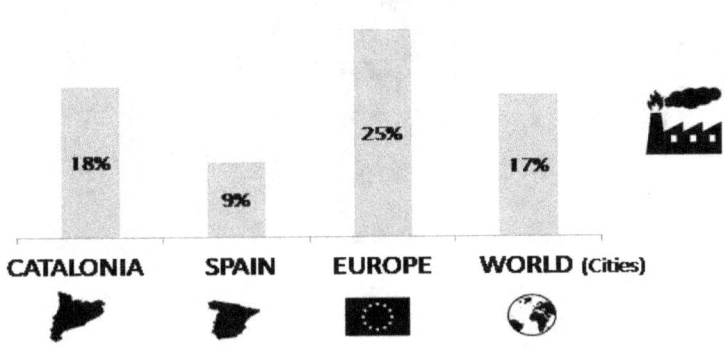

Of the total waste generated in 2012, in Catalonia, 16% was incinerated; in Spain, 10%; in Europe, 25% (over double), and in the main cities around the world, 17%. Several incineration facilities are available to treat all of this waste: 4 in Catalonia;[45] a total of 10 in Spain;[46] and more than 400 throughout Europe.[47]

The main arguments for incineration focus on the safety of these facilities, energy recovery in an environment of global energy crisis, the reduction of emissions that cause climate change, the low occupancy of public land and the improvement of technology to reduce the pollution generated:

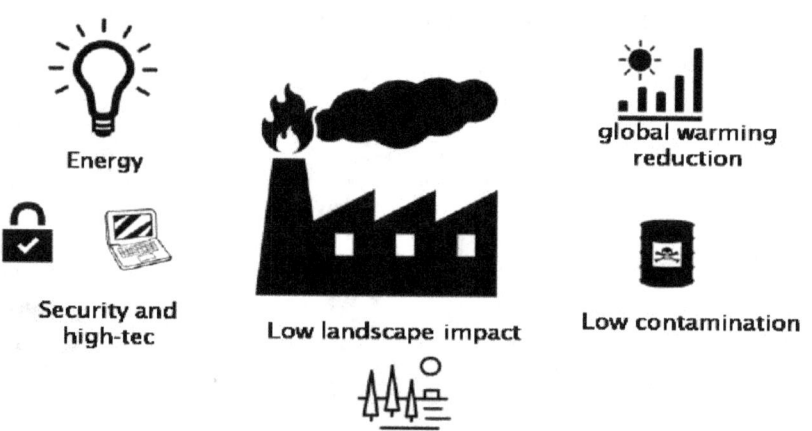

Incinerators are modern facilities that enable energy recovery with little impact on the landscape

45. PINFRECAT :Territorial plan for municipal waste management infrastructures in Catalonia 2013-2020
46. Annual report 2013 Ministerio de Agricultura, Alimentación y Medio Ambiente, Spanish Goverment
47. "Energy recovery, a necessary link?" www.laboratorioderesiduos.es

In an environment of global energy crisis, which depends on a limited resource like oil, one must ask oneself whether burying waste that has an energy value in landfills is acceptable or not. It's clear that the **energy needed to produce new products**[48] (which will become waste) **is higher than that which can be extracted from the product material through incineration**. Therefore, recycling is the best way to recover energy from these materials, as long as it's viable.

That said, if the waste is mixed (as in the rejected fraction), once it has passed through the Ecopark, it can no longer be recovered, because the energy (and economic) costs of recovery increase and the balance is no longer favorable, so the only option is to send this rejected material to an incinerator or to landfill. Adding energy from non-recyclable municipal, industrial, forestry and livestock waste, in Spain, could save at least 8% of the energy[49] consumed in one year.

Recycling is the best way to recover energy from waste

The emission of dioxins is no longer a problem[50] thanks to the gas purification systems featured in the current incinerators. According to the German Ministry for the Environment, between 1990 and 2000, emissions

---
48. "The incineration and future of waste management policies," Official Association of Industrial Engineers of Catalonia, March 2009
49. Alvaro Feliu, Luis Otero, "Eco-efficient waste recovery. Its potential in Spain," *Gas Natural Foundation*, 2007
50. "The incineration and future of waste management policies," op. cit.

from waste incineration facilities in Germany were reduced by a factor of almost 1,000 times and currently constitute less than 1% of emissions produced as a result of human activity. In the same vein, and according to several studies[51] by AEVERSU (Spanish Association of Energy Valorization of municipal solid waste) carried out in the surroundings of several of the Spanish incineration facilities such as Reus, Mataró, Tarragona or Zabalgarbi in Bilbao, these types of waste treatment facilities neither have an impact on their environment nor on human health. Equally, the Basque government[52] department affirms that there is no scientific evidence to suggest that modern incineration with limited emission levels implies a significant additional risk to the health of the population.

It should be noted that incinerators are municipal property, and are regulated, controlled and legislated so that they do not pose any risk to the environment or to people.[53] What's more, today's **incineration facilities** are not just incineration facilities; they **are high-tech power facilities.**[54]

Incinerators occupy less land than landfills and reduce the volume of waste that goes into landfills (around 85%).

---

51. Health and environment. www.aeversu.org
52. A study by Osakidtza concludes that incineration does not represent a "significant risk" to health. Guipuzkoa News, 7 March 2016
53. Catalan waste Agency, www.residus.gencat.cat
54. "The incineration and future of waste management policies," Official Association of Industrial Engineers of Catalonia, March 2009

The main **arguments against** incinerators focus mainly on the drawbacks implied for the environment and people. These drawbacks are caused by the different types of waste generated in an incinerator: slag, ashes, and emissions (dioxins).

Dioxine

Global warming compared

Dangerous ashes

One of the negative aspects of incineration facilities that caught my attention and which has been written about most are the **dioxins**. Dioxins are highly toxic and can cause reproductive and developmental problems, damage the immune system, interfere with hormones and also cause cancer.[55] The *orange agent* used by the United States in Vietnam had a high content of dioxins.

Burning waste emits carcinogenic dioxins and furans

---
55. Dioxins and their effects on human health," WHO (World Health Organization), May 2014 http://www.who.int/mediacentre/factsheets/fs225/en/

In terms of energy production, incinerators generate electricity but emit 33% more greenhouse gases than thermal power facilities that produce electricity from gas.[56]

Although this has already been mentioned, it should be highlighted that **the potential for saving energy[57] from municipal waste is higher through its recycling** than through the energy extracted from said waste (this is important).

Nor should we forget that waste incineration, like landfills, has associated fees or taxes, meaning that everything that is incinerated is around 20% more expensive[58] for citizens, compared to recycling (see chapter 10). What's more, these rates do not include all of the environmental costs[59] caused by incineration.

It should also be noted that waste incineration in cement corporations is carried out by private companies that, due to their very nature as companies, prioritize economic matters over the impact generated by their activity on the environment and people.

Waste incineration is not the solution for garbage treatment

---

56. "A Changing Climate for Energy from waste, for friends of earth," Dominic Hogg report for the well-known consultancy Eumonia
57. Incineration is not the solution" Greenpeace Spain, www.greenpeace.org/espana/es/
58. The incineration of waste in figures. Socio-economic analysis of the incineration of municipal waste in Spain," Greenpeace, July 2010
59. *Ibidem*

As a personal conclusion, I understand that incineration facilities are necessary when waste is not recycled or can't be separated. It's true that they have evolved a lot and that facilities nowadays are much safer than those of previous times, but there are still some serious potential risks and uncertainties that indicate that their general implementation is not recommended. **The incineration model is not the solution to waste treatment.** Incinerators should be reserved for treating waste that can no longer be recycled or where separation costs are excessively high and, as a precaution, placed in unpopulated areas. Thus, incinerators would be a last resort, although they are better than landfills.

Incinerators occupy the penultimate position in waste management hierarchy. First, you should reduce; if that's not possible then you should reuse; and if you can't reuse it either, you should recycle it; and if nothing else can be done, then yes, send it to for incineration. Landfill should be last option, but before that you have incineration. It's important to insist that incineration is the penultimate of the possibilities and not the priority or general model for massive waste incineration. No. First, you should recover the waste materials as much as possible internally, and if this is impossible, recover their energy. **From the energy point of view, it's better to recycle than to incinerate.**

Want to know more?

▶ Sant Adrià incineration facility (MUSICAL)

https://www.youtube.com/watch?v=yRVMNX8sXss&list=PLWQMeO43vsuf_k7ScZM_CpaJzf-tw7GEY&index=6

▶ Interview with Dr. Eduard Rodríguez Farré at the Vilafranca del Penedès Regional Hospital (CAT)

https://www.youtube.com/watch?v=StvcusDajGw

▶Incineración and health. Osasuna eta errausketa. April 2016. Legazpi (CAST)

https://irabaziganarlegazpi.wordpress.com/2016/04/28/incineracion-y-salud-osasuna-eta-errausketa-abril-2016-legazpi/

▶Award-winning documentary *TRASHED*, starring Jeremy Irons. Very instructive, easy to understand, with good images, good content and good technical details (ENG)

http://www.trashedfilm.com/

▶ VI State meeting against the incineration of waste in cement factories. Villafranca del Penedés 2015 (CAST and CAT)

https://vimeo.com/122267601

▶ "Purifying fire?" *El Escarabajo Verde* Program, RTVE, (CAST)

https://www.youtube.com/watch?v=qSL2NI8BHKU

▶ On the case "Txingudi incinerator (Basque Country)" (CAST)

https://www.youtube.com/watch?v=B4E9G5khk5c

i Effects on health and the environment of dioxins and furans, Spanish Ministry of Agriculture, Food and Environment

http://www.prtr-es.es/Dioxinas-y-Furanos-PCDDPCDF,15634,11,2007.html

▶ The reality of living near an incinerator, RTVE (CAST)

https://www.youtube.com/watch?v=IseUTKryYRM&feature=youtu.be

i Several studies on the non-impact of incineration on the environment and humans, by AEVERSU (Spanish Association for the Energy Recovery of RSU) (CAST)

http://www.aeversu.org/index.php/es/valorizacion-energetica/salud-y-medio-ambiente

i "The activity of the urban waste incineration does not imply an additional risk for the sourranding population." Universitat Rovira i Virgili (CAT)

http://wwwa.urv.cat/noticies/diari_digital/cgi/principal.pl?fitxer=noticies/noticia017204.htm

i Information websites

http://noalaincineracion.org/por-que-no/

http://www.contraincineracio.org/

6.
RECYCLING IS THE SOLUTION

"Recycle" comes, etymologically speaking, from the Greek word kýklos, which means "orbit or circle," and by extension, "ordered repetition or recurrence of phenomena." In Latin, the term becomes *cyclus-cycli*, adding the prefix: "Re." (which means repetition). Therefore, the original meaning of the verb "recycle" is "to circulate something or put it back into orbit" (waste becomes resource).

Garbage or waste recycling has been practiced for years, if not centuries or millennia: thousands of years ago, metal was repeatedly melted to forge new objects or products, such as weapons or tools. It is even said that the broken bronze pieces of the Colossus of Rhodes[60] (one of the seven wonders of the Ancient World) were recycled as scrap metal.

---
60. "The thruth about recycling," *The Economist*, 7 June 2007

So far, we have covered the vast amount of waste which is generated on different scales, and how traditional waste treatments, be they landfill or incineration, continue to have an effect on the environment (pollution and climate change) and how, on the other hand, they are not the global solution to the waste problem.

Waste recycling is a millennial practice

Considering that most of the waste could be recycled, the next alternative could be to RECYCLE the waste. When I refer to recycling, I do so in a **broad sense**, that is: **recycling waste, reusing it, composting it and also reducing** the amount of waste that we generate (prevention).

The waste recycling process includes the sub-processes of separation, collection, and waste treatment to obtain materials that can be used to make new products.

An alternative title for this book could have been: "Why it's important to reduce waste." However, if we can only recycle 40% of waste (by collecting it separately into containers) and the whole population is not collaborating, how can we ask people not only to separate their waste into different containers but also to avoid generating waste in the first place? Therefore, let's start by explaining the basic notions about recycling and why it's so important to do so, although I must reiterate that **it's preferable to reduce** the amount of waste that we generate.

The initial 3R strategy (reduce, reuse and recycle) has evolved towards a modern waste management system that establishes guideline principles or a **waste hierarchy** (European Union[61]) to find out what to do with waste:

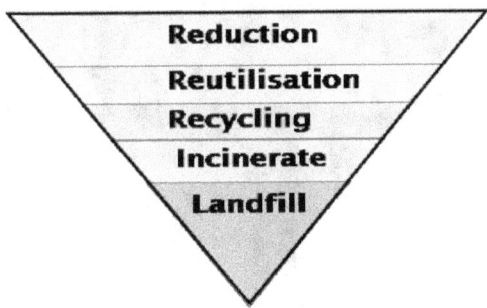

What to do with the waste?
Reduce, reuse, recycle, incinerate and, as a last resort, send to landfill

---

61. Directive 2008/98/CE European Parliament, about Waste Management, 19 November 2008

Stop garbage

The first option in waste management is *reduction* (prevention). The aim here is to avoid generating waste in the first place, and not only in terms of the quantity but also its level of danger or toxicity. If said generation can't be prevented, the second step in the strategy is to *reuse* (or prepare for reuse) the waste that we generate, such as furniture, clothing, appliances, etc. If neither of the first two steps is possible, waste should be *recycled*, as materials, to obtain new objects. If recycling is not feasible, then the waste must be recovered, at least, for example through *energy recovery*, which includes incineration. The last, least desirable step, and only when the previous options are impossible, is to send the waste to *landfill*.

Under all circumstances, waste recycling ranks above the alternatives of incineration and landfill. The level of energy needed to make products[62] is higher than the energy that can be extracted from the waste material by means of incineration. Also, concerning climate change, **recycling is better than incinerating.**[63]

---

62. "Incineration and the future of waste management policies," Official Association of Industrial Engineers of Catalonia, March 2009
63. Dominic Hogg. op.cit

Recycling is the solution

The waste management model that generates the least amount of greenhouse gas emissions[64] is municipal waste separation at the origin –at home– enabling recycling (and compost), combined with general waste treatment in Ecoparks (to treat the biodegradable waste fraction which is not separated).

Recycling waste is better than sending it to landfill or for incineration

One must take into account that the activity itself and the recycling[65] industry also have impacts on the environment (waste transport, recycling facilities operation), although in most cases the effects avoided are more significant than those generated by the activity of recycling.

Waste separation at home (at the origin), into materials or fractions, is important in order to ensure the maximum quality of the recycled materials since it helps to increase their value and the number of products that can be made.

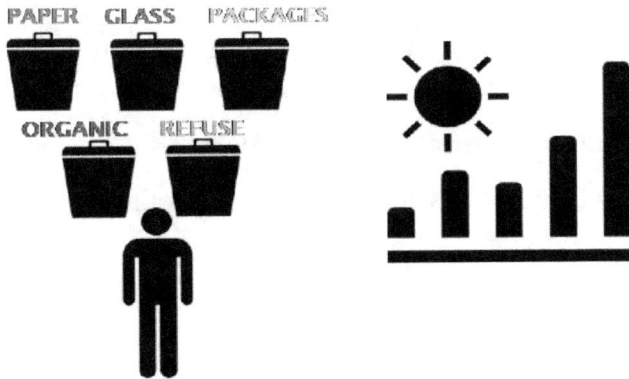

---

64. "Waste management options and climate change," European Commission, AEA Technology
65. "Environment and Europe: Estate and perspectives" European Environment Agency, 2010

In the next chapters we'll look at the main reasons why it's important to RECYCLE (reuse, recycle and compost) the different materials such as plastic, glass, paper, etc., but I'll summarize this for you first:

- To reduce greenhouse gases, which contribute to climate change by heating our planet, the Earth
- To reduce energy consumption, especially in countries with supply difficulties. The energy needed to make products[66] is higher than the energy that can be extracted from the waste material through incineration
- To reduce consumption of materials obtained from the soil, which are finite, such as metals, wood, fertilizers, water, precious materials, among others. Above all, in Europe, since its industry depends on the importation of raw material for manufacturing
- To reduce air pollution: gases that are harmful to our health and pollutants that worsen the quality of the air that we breathe
- To reduce soil and water contamination, for example, in uncontrolled dumps, which end up filtering into the subsoil
- To reduce the waste that ends up in landfill or is incinerated. Recycling comes first in the waste hierarchy
- Recycling generates employment
- Recycling is cheaper than not recycling
- Waste that was previously discarded turns into a resource

---

66. "La incineración y el futuro de las políticas de gestión de residuos" *op.cit.*

Bearing these benefits in mind, which are all very desirable, in the following chapters we will detail, quantify and explain what exactly is the impact on the environment, society, and people's health.

7.
HOW IMPORTANT IS RECYCLING?

7.1. Glass: a successful case study

Glass is an inorganic material that is obtained mainly from silica sand (SiO_2, around 70%[67]), calcium carbonate (Ca2CO3, around 15%), limestone ($CaCO_3$, approximately 10%) and other additives, by melting at about 1,500 °C (2,732 °F).

According to different sources, glass was first manufactured around the year 3500 a.C. in Egypt -other sources indicate its origin to be 2500 a.C. in Mesopotamia[68]- as a decorative element, and its different uses have evolved. Thus, glass has been used to manufacture vessels, windows and nowadays, above all, for packaging, specifically for liquids, since it's a type of packaging that neither interferes nor alters the taste of its contents. We have been manufacturing glass for the last 4,500 years, at least.

---

67. "Recycler le verre d'emballage. Porquoi?," Verre Avenir, Chambre Syndicale des Verreries Mécaniques de France
68. *Ibidem*

Glass is a material that can be fully recycled in an unlimited manner, meaning that you can repeat the process as many times as you want without altering its properties.

The color of the glass depends on the type of additives that are put in. You can usually find it in 3 different colors: green, topaz or amber and transparent. Separation or triage by colors can be carried out before starting the recycling process. However, this triage is optional. In facilities such as Barcelona, for example, the bottles are usually crushed or fragmented with the 3 types or colors mixed together. The crushed material is called cullet, which is taken to the bottling factories to be melted in an oven in order to obtain new glass bottles. 60-70% of the glass bottles on the market are green – wine, cava, champagne. Some factories use up to 95% of cullet to manufacture new glass bottles.[69]

> The glass is crushed, melted and reused to make glass bottles. The process can be repeated infinitely

In countries like Germany or Ireland, glass is separated by colors. Other countries have even chosen to climb one step higher on the recycling pyramid that we saw in the previous chapter, and instead of recycling glass, they reuse the glass packaging through a deposit-refund system (or DRS).

---

69. *Ibidem*

How important is recycling?

Around **8%** of the household and commercial garbage that we generate is glass.[70]

In 2012, a total of 169,222 tons of glass (the equivalent of 646 million bottles) were recycled (collected separately in green containers) in Catalonia.[71] In Spain,[72] 726,729 tons (2,774 million bottles) and in Europe,[73] 15,700,000 tons (almost 60 billion bottles).

Of the total amount of glass generated in 2012 in Catalonia,[74] **70%** was recycled (recovery, which includes glass recovered from Ecoparks), along with **69%** of the glass generated in Spain[75] and **73%** in Europe. There's no doubt about it, this is excellent recycling data, especially if we compare it with the other fractions, as we will see afterwards.

	CATALONIA	SPAIN	EUROPE
GLASS recycled (Tn/year)	169,222	726,729	15,700,000
BOTTLES recycled (millions/year)	646	2,774	59,924

70% of the glass produced is recycled.
A success story that began in 1982

---

70. My compilation from different studies: op. cit.
71. Generalitat de Catalunya Statistics www.estadistiques.arc.cat
72. Annual report 2013 Ministerio de Agricultura, Alimentación y Medio Ambiente, Spanish Government
73. EUROSTAT. Official statistics. European Union
74. My compilation from different studies. Official data for Catalonia is 64% but from General prevention and waste management Program in Catalonia 2013-2020
75. EUROSTAT. Official statistics. European Union

Judging from these numbers, we could say that the history of glass recycling is a success story that has not yet come to an end and that continues to thrive. Little wonder, as the selective glass container was the first one to be placed on the streets, back in 1982, as part of the first recycling campaign in Barcelona.

BENEFITS OF RECYCLING GLASS

The main benefits of glass recycling are:

- **Energy saving:** Using crushed glass from bottles (cullet) instead of virgin material saves between 20 and 30%[76] of energy. The key point to this energy saving is that, when manufacturing is carried out using recycled glass or cullet, the material's **melting point is lower** and subsequently, so is the energy required. (Recycling one bottle is equal to the energy consumption of 1 110-watt bulb for 4 hours; recycling 3 bottles to the use of a dishwasher service and 4 bottles to a refrigerator for one day[77])

- **Savings in greenhouse gas emissions** (GHG). By using glass cullet, the carbon emissions are reduced from between 20[78] to 50%.[79] For every recycled ton of glass, the emission[80] of 200 kg of CO_2 (166[81]-315[82]) is avoided

---

76. "Glass Recycling Facts," Glass Packaging Institute. And also the same in, Alejandro Mata and Carlos Gálvez, "Conocimiento del proceso de reciclaje de envases de vidrio; propuestas de mejora del proceso actual y análisis costo-beneficio de la implantación del mismo en la planta Vidriera Guadalajara," Universidad Autónoma de Guadalajara
77. Benefits. www.ecovidrio.es
78. Carbon Footprints.. www.o-i.com
79. Every 10% of cullet used reduces 5% of CO2. Source: "Recycler le verre d'emballage. Porquoi?," Verre Avenir, ChambreSyndicale des VerreriesMécaniques de France
80. Recycler le verre d'emballage. Porquoi? Verre Avenir. ChambreSyndicale des Verreries Mécaniques de France
81. "Glass Recycling Facts," Glass Packaging Institute
82. "Glass recycling information sheet," www.wasteonline.org.uk.

- **Improvement of air and water quality** by reducing its pollution: air pollution is reduced by 20%.[83] For every 10%[84] of recycled glass, the emission of particles into the atmosphere is reduced by 8%, 10% in the case of sulfur oxides and 4% for nitrogen oxide (responsible for air pollution in cities such as Barcelona or Madrid)

- **Saving of raw materials** and conservation of the environment as the need to extract is reduced: For every 1kg of glass bottles made from cullet, 1.2 kg[85] of virgin materials (sand, limestone and sodium carbonate) is saved

- **Saving resources**: By recycling, the glass manufacturing industry could be supplied with almost 34% of the resources that it needs[86]

- **Improvement of water quality** by reducing pollution between 40[87] and 50%[88]

- **Less waste is sent to landfill**: For every 3,000 recycled glass bottles, 1,000 kg[89] of garbage is prevented from ending up in landfill

Glass recycling saves energy, resources, GHG emissions and improves air quality

---

83. www.panda.org and www.ecovidrio.es
84. "Glass Recycling Facts" .*op.cit*
85. Carbon Footprints.. O-I.com and Ecovidrio .*op.cit.*
86. General prevention and waste management Program in Catalonia 2013-2020
87. Alejandro Mata and Carlos Gálvez, *op.cit.*
88. www.panda.org
89. www.ecovidrio.es

Want to know more?

▶ Recycling glass. Agència de Residus de Catalunya (MUSICAL)

https://www.youtube.com/watch?v=tFCCdhvaldE&list=PLWQMeO43vsuf_k7ScZM_CpaJzf-tw7GEY

🛢 Recycling data in different regions in Spain (CAST)

http://www.ecovidrio.es/reciclado/datos-de-reciclado/estadisticas

🛢 Information about the DRS system (Deposit Refund System) (ENG and CAST)

https://www.surfrider.eu

http://www.retorna.org

🛢 Plastic bottle deposit return scheme could save England's councils £35m a year (ENG)

https://www.theguardian.com/environment/2017/oct/11/plastic-bottle-deposit-return-scheme-could-save-englands-councils-35m-a-year

7.2. Paper comes from trees

Paper is an organic material that is primarily made from the cellulose fibers of virgin wood from trees which is used to obtain pulp. This cellulose pulp can be obtained from virgin wood or also from recycled paper.

The primary woods used to manufacture said cellulose pulp are called "pulpable woods," mostly softwood trees such as spruce, pine, fir or larch, although hardwood from fast-growing trees such as eucalyptus or birch is also used, or eucalyptus and pine in Spain.[90]

Once the cellulose pulp or paper pulp is obtained, it goes through several mechanical processes, the fibers are separated -they're joined by a kind of glue called "lignin"-, mixed with water and then dried to obtain the paper reel. A bleaching process is usually applied to paper.

Paper production[91] accounts for approximately 35% of tree felling around the world.

---
90. www.aspapel.es
91. Mjnsbzgkxartin, Sam (2004). Paper Chase. Ecology Communications, Inc.. Retrieved 2007-09-21

Cardboard is made up of several layers of paper, which are superposed and glued from virgin material or recycled paper. Cardboard is thicker, harder and more resistant than paper. Most cardboard boxes are made using a corrugated cardboard structure with smooth layers and corrugated layers on the inside, improving mechanical characteristics and increasing resistance during transport and storage. Pinewood is the raw material most used in cardboard production.

> 35% of trees felled are destined for paper manufacturing

Throughout history, many different writing supports were used before the discovery of cellulose paper. In ancient Egypt, in 2000 BC, papyrus was used. In China,[92] in the year 105, the first paper was produced from scrap silk, rice or hemp. In Europe, in the Middle Ages, scrolls were made from goat skins or tanned ram hides, and later, in the fourteenth century, paper was made out of cotton. The production of paper using cellulose didn't come about until the eighteenth century, with the implementation of the Kraft process.

The writing support has evolved to such a point that, nowadays, we have digital and media supports that allow us to read and write on many different devices such as e-readers, computers, tablets or smartphones.

---

92. "The paper, the main actor of our history," Aspapel

Around **12%** of household and commercial waste is paper and cardboard.[93]

In 2012, in Catalonia,[94] a total of 318,210 tons of paper-cardboard were recycled (collected separately into containers), in Spain,[95] the figure was 1,085,574 tons.

It is estimated that, in 2012,[96] recycling (recovery, which includes paper recovered in Ecoparks) made up 46% of the paper-cardboard (of household sand businesses) generated in Catalonia. There is no verified data from Spain or Europe, as the data includes paper recycled by industries.

46% of the paper produced is recycled. Recycling can be repeated up to 6 times

Unlike the infinite recycling of glass, paper can be recycled, on average, about 6 times.[97] This is because its pulp fibers become cut and frayed and there comes a time when they are so small that they lose their consistency and can no longer be recycled. For this reason, every time paper is made, virgin fiber must be added to ensure good quality. However, the only difference between recycled and virgin fiber is that each one is in a different phase of its *life*.

---

93. My compilation from different studies: "Pesa la brossa" 2014. Study for the Polytechnic University of Catalonia and General Program of Prevention and Waste Management of Catalonia 2013-2020. According to the Agència de Residus de Catalunya 2014, the data are Organic 37%, paper, and cardboard 12%, glass 8%, plastics and metals 12%. "La gestió dels residus i el seu impacte en el canvi climàtic." Statistics 2014
94. Generalitat de Catalunya Statistics www.estadistiques.arc.cat
95. Annual report 2013 Ministerio de Agricultura, Alimentación y Medio Ambiente, Spanish Government
96. General prevention and waste management Program in Catalonia 2013-2020
97. Source Aspapel, interview in BLOG *El País* "How many times could be recycled?"

For responsible paper consumption, it's advisable to use it on both sides and buy recycled paper (non-chlorinated). If this isn't possible, at least make sure that it carries a certification proving that it's the result of controlled tree felling which is managed sustainably, such as the popular NGO FSC (Forest Stewardship Council) certification.

BENEFITS OF RECYCLING PAPER

Let's summarize the main benefits of paper recycling:

- **Energy saving**: Paper produced from recycled paper represents an energy saving of 70%[98] compared to the energy needed to produce paper from wood or virgin fibers
- **Reduction of the raw material consumed** (trees felled): For every ton of recycled paper, the equivalent of 12 trees[99] (4m^3 of wood) is saved in wood. Other sources indicate 17 trees[100] and even[101] 31
- **Saving resources**: By recycling, the paper-cardboard industry could be supplied with almost 69% of the resources[102] it needs
- Water saving: Recycling paper saves 80% of water compared to production from virgin fiber
- Improved air and water quality: Paper recycling constitutes a 74% reduction in gas emissions and a 35% reduction in water[103] polluting emissions
- **Savings in greenhouse gas (GHG) emissions**
- **Less waste is sent to landfill or for incineration**

---

98. Ministerio de Agricultura, Alimentación y Medio Ambiente
99. *Ibidem*
100. Bureau of International Recycling. Website
101. *Ibidem*
102. General prevention and waste management Program in Catalonia 2013-2020
103. Ministerio de Agricultura, Alimentación y Medio Ambiente and Bureau of International Recycling

Paper recycling saves a lot of energy and water, reduces tree felling (12 per ton), GHG emissions and improves air quality

Per 1 ton (1,000 kg) of recycled paper:

- The felling of at least 12 trees is avoided. **If one person recycles all of the paper that they produce during one year, they save almost one tree from being cut down**
- 4,000 KWh[104] of energy is saved
- It implies a saving of 26 m³ in water
- The equivalent of 3.5 m3 in landfill space
- In paper[105] manufacturing, each time 1 ton of virgin fibers is replaced by recycled paper and cardboard, 2.3 tons of CO_2 equivalent are saved, which would cover a distance of 13,501 km. Carbon dioxide (CO_2) is one of the greenhouse gases that causes climate change
- 0.9 tons of paper[106] can be produced
- Recycling 8 cardboard cereal boxes could make a book[107]

---

104. Bureau of International Recycling
105. "Best practices in paper and cardboard recycling in Catalonia," Agencia Residus de la Generalitat de Catalunya and Gremi de Recuperació de Catalunya
106. *Pourquoi trier les ordures*, Mairie du Paris
107. Ecoembes, Equivalence and data, www.ecoembes.es

Want to know more?

▶ Video: Paper manufacturing process (ENG)

https://www.recyclenow.com/what-to-do-with/paper-0

▶ Video: Paper recycling process (CAST)

https://www.youtube.com/watch?v=Rc_MsY6s-nA

i Greenpeace complaint about excessive plantations of Eucalyptus in China (ENG)

http://www.greenpeace.org/eastasia/campaigns/forests/problems/china-remaining-forests/

i How to reduce consumption and optimize the use and recycling of paper (CAST)

http://www.greenpeace.org/espana/Global/espana/report/other/el-papel.pdf

i Guide of good practices for the recycling of cardboard and paper by the Generalitat de Catalunya (CAT)

http://residus.gencat.cat/web/.content/home/lagencia/publicacions/centre_catala_del_reciclatge__ccr/cast_guiapapercartro_web.pdf

i More information about the FSC certificate (ENG)

https://ic.fsc.org/en

7.3. The importance of recycling plastic and metal (packaging)

Packaging is usually used to contain, protect, handle or deliver goods like food or products in general. Besides the content, packaging also includes auxiliary or ancillary elements (such as labels, lids (on yogurt pots, for example), fillings, among others).

Packaging can be made from different types of material, such as glass, cardboard, plastic or metal, among others.

The yellow container (color in Spain and generally in Europe, too) is for packaging waste (this fraction does not include paper-cardboard packaging or glass packaging).

About **12%** of domestic garbage that we generate in our homes and stores is plastic and metal.[108] This packaging waste fraction includes:[109]

- **Plastic** packaging, approximately 50% of packaging[110]
- **Metal** packaging (steel or aluminum), about 33%
- **Composite or mixed** packaging makes up 17% of packaging. Among these are cartons or Tetra Briks,[111] 74% of which is paper-cardboard, 21% plastic and 5% metal (tetrapak).

---

108. My compilation from different studies: "Pesa la brossa" 2014. Study for the Polytechnic University of Catalonia and General Program of Prevention and Waste Management of Catalonia 2013-2020. According to the Agència de Residus de Catalunya 2014, the data are Organic 37%, paper, and cardboard 12%, glass 8%, plastics and metals 12%. "La gestió dels residus i el seu impacte en el canvi climàtic." Statistics 2014

109. This fraction does not include paintor chemical product packaging. This type of waste should be taken to a reuse and recycling center

110. Agència de Residus de la Generalitat de Catalunya residus.gencat.cat

111."Carbon footprint and municipal waste management in Catalonia," INEDIT and Agencia de Residus de la Generalitat, 2011-2012

To be as brief as possible in this chapter, I'll focus on plastic and metal packaging, since they are the most common types.

In 2012, 135,378 tons of packaging waste was recycled (separately collected into bins) in Catalonia.[112] In Spain,[113] this figure was 641,266 tons.

In Catalonia, it is estimated[114] that the recycling rate (**homes and businesses**) **was 30%** of the total packaging waste generated in Catalonia (including packaging recovered in Ecoparks). In Europe,[115] the plastic packaging recycling rate (excluding metals) was 35% in 2012. In the same year in the USA,[116] 28% of PEAD plastic packaging and 31% of PET packaging was recycled (acronyms defined in the following pages). There are no verified figures for Spain since the data includes packaging recycled by industries.

30% of the municipal plastic and metal packaging produced is recycled[117]

---

112. Generalitat de Catalunya, Statistics estadistiques.arc.cat
113. Annual report 2013 Ministerio de Agricultura, Alimentación y Medio Ambiente, Spanish Government
114. General prevention and waste management Program in Catalonia 2013-2020
115. 2013 facts from the European association of European plastic manufacturers, Plastic packaging waste in EU27, which was 40% in 2014according to the other source: "The new plastics economy. Rethinking the plastics economy," Ellen MacArthur Foundation, January 2016
116. The United States Environmental Protection Agency www.epa.gov
117. Ecoembes says that 70.3% of plastic, bricks, cans, and paper and cardboard packaging is recycled, year 2012. "Recycling packaging: present past and future."

PLASTIC PACKAGING (BOTTLES)

There is a wide variety of plastics. Most of them are artificial materials, although there are some which are natural. In the beginning, materials were considered plastic not because of their composition but because of their plasticity, their ability to take on different forms at certain temperature ranges. Plastics, in general, come from petroleum and other natural substances and are obtained synthetically by multiplying the carbon atoms present in their molecules (polymerization). Plastics are often light – or low in density–, are relatively cheap and very durable over time. Plastic is extremely integrated in the modern way of life.

Plastics are petroleum derivatives. They have only been in general use for 100 years and have grown exponentially

The history of plastic began in 1860 and 1909 marked the beginning of the "plastic era" with the manufacturing of the first fully synthetic plastic called Bakelite. Therefore, unlike other materials such as paper or glass, plastic has only been in use for about 100 years. However, use of plastic has grown exponentially and is expected to

Stop garbage

continue. Compared to the almost 2 tons[118] produced in 1950, this figure rose to 300 million tons in 2013, and it is expected that, at this growth rate, in 2050 world plastic production could triple.[119] It is estimated[120] that 26% of global plastic production is destined for the production of plastic packaging (other sources[121] indicate 40%).

It is calculated[122] that, in 2012, world plastic production reached almost 300 million tons. Plastic production accounts for between 6%[123] and 8%[124] of global oil consumption (half as raw material and the other half as energy required to manufacture plastic). Remember that oil is a finite resource.

118."New Link in the food cain? Marine plastic pollution and seafood safety" Web Environmental Health Perspectives ehp.niehs.nih.gov
119. Wurpel G.,Van den Akker J., Pors J., Ten Wolde, *Plastics do not belong in the ocean. Towards a roadmap for a clean North Sea.* IMSA Amsterdam,2011), p. 39
120. The new plastics economy. Rethinking the plastics economy, Ellen MacArthur Foundation, January 2016
121.Libro Verde: *sobre una estrategia europea frente a los residuos de plásticos en el medio ambiente,* European Comission
122. "Global Plastic Production Rises, Recycling Lags," World watch Institute.
123. "The new plastics economy. Rethinking the plastics economy," Ellen MacArthur Foundation, January 2016
124. "Global Plastic Production Rises, Recycling Lags," World watch Institute.

The main types of plastic packaging found in household and businesses garbage are:

• **PET** or Polyethylene Terephthalate is a very resistant and light material. It's usually used for the manufacturing of **water bottles** or carbonated soft drinks. It represents 15% of European plastic packaging (production)[125]

• **HDPE** or high-density polyethylene is a plastic which is resistant both to impacts and to low temperatures, it's waterproof and is an electrical insulator. It is used for **milk bottles or cleaning products and detergents.** It represents[126] 19% of European plastic packaging

• **LDPE** or low-density polyethylene is a soft plastic which is flexible and not terribly resistant to temperature that is used to make **plastic bags** like the ones we can find in supermarkets or shops, garbage bags or plastic film. It represents 32% of European plastic packaging[127]

• Other plastics such as PP or polypropylene or PS or polystyrene and expanded polystyrene (porexpan)

According to European law, all packaging must include identification or labeling indicating its composition, to facilitate collection and recycling.

---

125. "Plastic waste in the environment," European Comission
126. *Ibidem*
127. *Ibidem*

The plastic packaging from the yellow bins can be fully[128] recycled, 100%. It is estimated that the plastic recycling process can be repeated 4 or 5 times.[129]

Plastic packaging can be 100% recycled up to 5 times

The packaging recycling process starts in a triage facility where items are separated by material type and then taken to the treatment facilities. To carry out this selection, in Catalonia[130] there are 13 triage facilities and a total of 94 in Spain.[131]

Once in the treatment facility, the plastic packaging is crushed down into small plastic flakes. These flakes are melted into pellets to manage their properties and to obtain a suitable material from which recycled products can be produced. Some examples of recycled products that can be made are:[132] pipes (31%), industrial components (25%), bags and sheets (15%), garbage bags (10%), various (14%) (street furniture, hangers, footwear, etc.), bottles and jerry cans (3%), household goods (2%) or plastic packaging that is not destined for food products (such as detergents or household cleaning products).

---

128. Libro Verde: *sobre una estrategia europea frente a los residuos de plásticos en el medio ambiente*, Comisión Europea and same data Ciclopast, www.cicloplast.com
129. "How many times can we recycle?," *El País*, blog semanal. 2010
130. PINFRECAT: Territorial plan for municipal waste management infrastructures in Catalonia 2013-2020
131. Memoria 2013 del Ministerio de Agricultura, Alimentación y Medio Ambiente
132. "Plastic recycling," Ciclopast, www.cicloplast.com

METAL PACKAGING

Metals are materials or chemical elements that are characterized by their capacity to conduct or transmit heat and electricity. They are usually quite heavy or dense, and are generally solid at room temperature. Most elements on the periodic table are metals.

The history of metal use goes back to Prehistory, in particular to the Bronze Age (3500 a.C), which was followed by the Iron Age (1400 a.C.). Therefore, we have been using metals for over 5,500 years.

The main metal packaging found in the garbage bags of homes and businesses are:

- Steel cans (98% iron and 2% carbon, approximately)
- Aluminum cans, for foodstuffs
- Aerosols (deodorants, cleaning products, etc.)

We have been using metals for over 5,000 years. This material can be endlessly recycled

Metals, like glass, can be endlessly recycled without altering their characteristics. Metallic packaging is 100% recyclable as melting processes allow a suitable material to be obtained which can be used to manufacture new products. Aluminum recycling produces a product which is practically the same as virgin aluminum.

In 2012, 41% of aluminum packaging was recycled[133] in Spain. It is estimated that 70% of the cans on the Spanish market are aluminum while steel cans make up 30%.

---

133. Association for the aluminum products recycling, www.aluminio.org

BENEFITS OF RECYCLING PLASTICS AND METALS

Here's a summary of the main benefits of its recycling:

- **Energy saving**: Production using recycled packaging represents an 84% energy[134] saving in plastics, 95% in aluminum cans and 75% in steel cans, compared to the energy needed for production from virgin materials.

- **Reduction in need for raw material**: For every ton of recycled plastic containers around 1 ton of oil[135] is saved. For every ton of recycled aluminum, you save 6 tons of bauxite (element from which aluminum is made)

- **Savings in resources**: All of the packaging recycled could supply the plastic manufacturing industry with almost 9% of the resources[136] that it needs

- **Improving air quality** by reducing pollution: Aluminum recycling represents a reduction in emissions[137] of 9.8 tons of CO_2 and 64kg SO_2

- **Savings in greenhouse gas (GHG) emissions**

- **Less waste is sent to landfill or for incineration**

Plastic and metal recycling saves a lot of energy, reduces the consumption of resources (oil) and GHG emissions as well as improving the air quality

---

134. Annual report 2013 Ministerio de Agricultura, Alimentación y Medio Ambiente, Spanish Government. And also Bureau of International Recycling
135. Ministerio de Agricultura, Alimentación y Medio Ambiente, Spanish Government
136. Prevention and waste management general program, Catalonia 2013-2020
137. Study on the recovery of aluminum, Arpal, (Association for aluminum products recycling), 2013

For 1 ton of packaging:

- You save 1 ton of **oil** (from one ton of plastic packaging)
- From PET bottles there's a net benefit in greenhouse gases of 1.5 tons of CO_2 equivalent[138]
- Recycling 1 ton of aluminum saves 6 tons of bauxite, 4 tons of chemical products and 14,000 kWh of **electricity**[139]
- Recycling **1 aluminum can** save enough energy to run a **TV for 3 hours**[140]
- Recycling **1 PET plastic bottle** saves enough energy to run a **TV for 20 minutes**[141]
- Recycling **40 bottles of water (PET)** can make a **polar fleece**[142]
- Recycling **80 cans of soda** can make a **bicycle tire**[143]

---

138. "Life out of plastic"
139. Metals - aluminium and steel recycling. In wasteonline.org.uk.
140. *Ibidem*
141. "Why it's a good idea to pay attention to the plastic you buy," *La Vanguardia* 18 February 2016
142. Ecoembes, Equivalence and data, www.ecoembes.es
143. *Ibidem*

Want to know more?

▶ History of a can (CAST)

https://www.youtube.com/watch?t=112&v=zeno6jQHzKU

▶ Packaging that leaves a mark "Envases que dejan huella," Escarabajo verde program, RTVE (CAST)

http://www.rtve.es/television/20150505/envases-dejan-huella/1138963.shtml

🛈 Information about the DRS system (Deposit Refund System) (ENG and CAST)

https://www.surfrider.eu/en/

http://www.retorna.org

🛈 Plastic-free or "Sin plástico": Cooperative that promotes plastic reduction (ENG)

https://www.sinplastico.com/en/

🛈 Plastics Europe web (ENG)

http://www.plasticseurope.es

🛈 The most common types of plastics (ENG)

https://www.cutplasticsheeting.co.uk/blog/uncategorized/the-5-most-common-plastics-their-everyday-uses/

7.3.1. The kidnapping of the yellow container

In waste management, many of the waste laws and programs that have a bearing on the collection and treatment model are based on two important principles:

Extended Producer Responsibility Principle

The producers[144] of goods (which later become waste) must pay the collection and treatment management costs of this type of garbage. For example, water bottling companies or yogurt producers should take charge of the collection and treatment of water or yogurt packaging, respectively.

Polluter Pays Principle

Those responsible for environmental pollution must assume the costs derived from said contamination. For waste, this principle is usually expressed as the extended producer responsibility scheme. This principle is also known as "Pay as you throw" (PAYT).

Polluter pays

To implement the Packaging Waste European Directive[145] and its principles, Spain decided to choose a *Collection and Recovery System* (CRS) for packaging producers (producers, packers, and suppliers) and to create a company to manage it: Ecoembes. Another of the possible options was the DRS model: Deposit, Refund System.

---

144. Directive 2008/98/ EC, on waste and repealing specific Directives, articles 8 and 15
145. European Parliament and Council Directive 94/62/EC of 20 December 1994 on packaging and packaging waste. The Directive refers to Return, Collection and Recovery systems. The Spanish system it is called, literal translation, as Integrated Management System (in Spanish Sistema Integrado de Gestión o SIG)

With the CRS model, a producer who wants to put a product on the market must pay a management fee (collection and treatment) to Ecoembes, although this rate is usually applied to the price of the product and, consequently, to the customer.

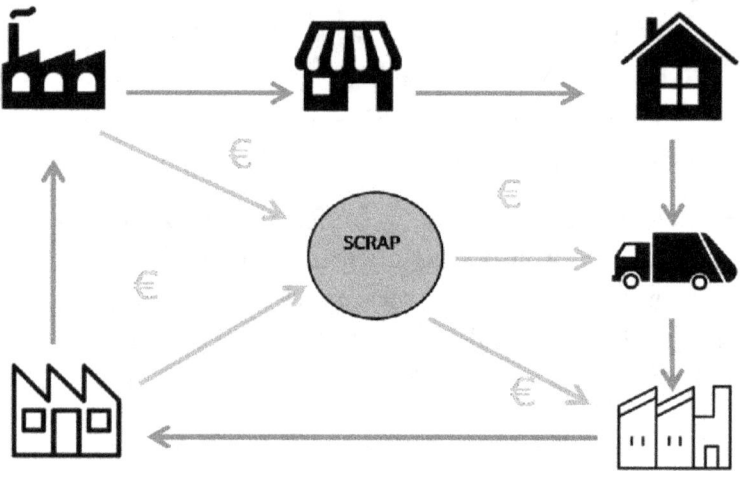

Ecoembes S.A. is a non-profit company but it has investee profit-oriented companies. In large numbers,[146] participation is as follows: 55% packaging companies (Nestlé, Pescanova, Coca-Cola, etc.), 20% commercial distribution companies (Corte Inglés, Carrefour, Mercadona) and 20% companies producing raw materials (Tetra Pack, Hispania SA, ARPAL ...).

---

146. "The packaging waste mortgage," magazine *El Ecologista*, web version N° 84, *Ecologistas en Acción*

On the other hand, to facilitate the collection of these containers, Ecoembes makes agreements or arrangements with the autonomic regions so that the municipalities can perform the collection (in a yellow container) and treatment of the packaging, in exchange for economic retribution. It's a kind of outsourcing, the objective being efficiency and preventing the cities from becoming overcrowded with different packing bin types and trucks.

Up to here, the model seems valid. However, at some point in the yellow container's implementation, it became associated with only packaging waste. Subsequently, no plastic or metal object/waste which is not packaging is recycled, even though it's the same material such as: toys, entertainment and sports goods, cleaning tools).

Analyzing or making characterizations[147] on generated waste, it could be find out that 43% of the plastics and metals that we throw into the garbage containers (42% only for plastic waste according to the EU) since they are not household packaging can not be put in the yellow container (they do not have the green recycling point that indicates that allows to deposite in it).

Waste treatment does not understand packaging. The packaging recycling facilities manage waste, such as the types mentioned before (PET, ALU, PP, PE, among others) and treatment is carried out per material. In fact, many non-packaging plastics are recycled with packaging, without any technical issues.

> Recycling does not understand packaging, only materials

Besides, the European Directives require waste management to be implemented separately, **by material type**, not by packaging. Let me quote, textually, Article 11 "Re-use and Recycling" from Directive 2008/98/EC on waste:

"Member States shall take measures to promote high quality recycling and, to this end, shall set up separate collections of waste where technically, environmentally and economically practicable and appropriate to meet the necessary quality standards for the relevant recycling sectors [...] by 2015 separate collection shall be set up for at least the following: paper, **metal, plastic** and glass."

---

147. My own compilation from waste characterizations data made in El Prat del Llobregat 2016, http://www.elprat.cat/actualitat/noticies/el-76-del-que-llencem-al-contenidor-gris-no-hi-hauria-danar

I'm not sure when this happened, but what's for sure is that packages kidnapped the yellow container. Fortunately, I think we can set it free ... without paying any ransom.

At some point in history, the yellow container became exclusively for packaging

PROPOSAL

I neither aim to be negative, nor to radically change the entire established recycling system... or maybe I do, and maybe we should switch to the DRS model, I have my doubts... However, we can do better: we could start by including any plastic or metal object –waste– in the yellow container, even if it's not packaging. By maintaining the current model, with a composition analysis of the waste in the yellow container (this is already currently done, it's called characterization), Ecoembes would only have to cover the cost of the packaging treatment. Why should we overload citizens with the responsibility and hassle of differentiating between what's packaging and what's not, knowing if this waste has paid the fee, or not? It would be easier to differentiate by materials (plastic and metal). Also, this way, other waste (non-packaging) would enter the recycling circuit optimally.

Want to know more?

🔖 Pay-as-you-throw system Guide (ENG)

http://residus.gencat.cat/web/.content/home/lagencia/publicacions/centre_catala_del_reciclatge_ccr/guia_pxg_en.pdf

🔖 Directive 2008/98/CE on waste (ENG)

http://eur-lex.europa.eu/legal-content/EN/TXT/HTML/?uri=CELEX:32008L0098&from=EN

🔖 Blog article: "Mortgage of packaging waste" (CAST)

http://www.productordesostenibilidad.es/2015/04/la-hipoteca-de-los-residuos-de-envases/

http://www.ecologistasenaccion.org/article29802.html

🔖 Plastic bottle deposit return scheme could save England's councils £35m a year, *The Guardian* (ENG)

https://www.theguardian.com/environment/2017/oct/11/plastic-bottle-deposit-return-scheme-could-save-englands-councils-35m-a-year

🔖 DRS proposal (CAST)

http://www.retorna.org/es/elsddr/propuesta.html

🔖 Study against DRS and supporting (RCS) (CAST)

http://www.envaseysociedad.org/analisis-de-solvencia-tecnica-de-diferentes-estudios-realizados-sobre-el-sddr/

🔖 Ecoembes (ENG)

https://www.ecoembes.com/en/home

7.3.2. The biggest plastic dump in the world: The ocean

New times, new challenges.

We have seen how plastic production has increased[148] to 300 million tons in 2013, of which 78 million[149] is destined for the production of plastic packaging. The extraordinary growth of this type of garbage has resulted in many plastics escaping collection and treatment systems and ending up in other places, like the seas or oceans.

> 32% of plastic produced ends up in the sea forming floating islands of waste

Every year, around 10 million tons of garbage[150] ends up in the seas and oceans. Most is plastic waste and, of this waste, plastic packaging represents[151] 32%. Plastic waste dumping is currently equivalent to **emptying a garbage truck into the sea every minute**. In the Atlantic and Pacific Oceans, there are floating islands[152] or waste patches which weigh around 100 million tons, of which 80% are plastics. **80% of marine plastics originate from land.** The main causes are river water discharges, hydraulic

---
148. "New Link in the food cain?," *op.cit.*
149. The new plastics economy. Rethinking the plastics economy.,Ellen MacArthur Foundation, January 2016
150.Libro verde: *sobre una estrategia europea frente a los residuos de plásticos en el medio ambiente,* European Comission. . Eight million found it in "The new plastics economy. Rethinking the plastics economy," *op.cit.*
151.The new plastics economy. Rethinking the plastics economy," *op.cit.*
152.Libro verde: *sobre una estrategia europea frente a los residuos de plásticos en el medio ambiente,* European Comission. . 150 million found it in "The new plastics economy. Rethinking the plastics economy," *op.cit.*

dam spills, tourism garbage, and industrial activities. If we are not able to stop this dumping and pollution rate, it is estimated[153] that there will be more waste plastics in the oceans than fish by the year 2050!

> Plastics affect the fish, and if we don't change our mentality, there will be more plastic in the sea than fish

Just think, the time required for nature to break down plastics can be hundreds of years.

The main problem or consequence of these floating plastics is, on the one hand, that marine species can get entangled in them and also eat them. If marine fauna ingests these plastics, especially microplastics (small particles of plastics decomposed by the sun) and chemical additives, this can lead to a significant potential source of contamination.[154] On the other hand, these plastics usually have toxic substances which may be incorporated into the environment or **water**.

The focal point of regional pollution is China, which bears 30% of the responsibility, followed by Indonesia, the Philippines, Vietnam, Sri Lanka, among others.[155]

---

153. "The new plastics economy. Rethinking the plastics economy," *op.cit.*
154. Libro verde, *op.cit.*
155. "Which countries create the most ocean trash?" Article *The Wall Steert Journal*

Want to know more?

▶ Video about plastics in the ocean (ENG)

http://oceantoday.noaa.gov/trashtalk_garbagepatch/

▶ A can inside a fish. Very clear and concise

https://www.youtube.com/watch?v=M6CfwOkDAeg

▶ Documentary: *Garbage island: A full ocean plastic* (ENG)

https://www.youtube.com/watch?v=D41rO7mL6zM

🕮 Marine pollution article. Environmental Health Perspectives (ENG)

http://ehp.niehs.nih.gov/123-a34/#r2

🕮 Tide of plastic rubbish discovered floating off idyllic Caribbean island coastline, *Independent* (ENG)

http://www.independent.co.uk/news/world/americas/plastic-rubbish-tide-caribbean-island-roatan-honduras-coast-pollution-a8017381.html

🕮 GREEN PAPER *On a European Strategy on Plastic Waste in the Environment* (ENG, FR, CAST)

http://eur-lex.europa.eu/legal-content/ES/TXT/?uri=CELEX:52013DC0123

🕮 Web about a Global Coalition Movement to make a free plastic world (ENG)

http://www.plasticpollutioncoalition.org

▶ Captain Charles Moore discoverer of floating garbage islands (ENG)

http://www.ted.com/talks/capt_charles_moore_on_the_seas_of_plastic?language=es#t-44995

7.4. Biowaste: A key point

Vegetable and food leftovers (mostly from vegetable or animal origin) are called the "organic fraction" or biowaste. Biowaste consists of water (80% of its weight) and organic matter (carbohydrates, proteins, and fats). This waste is generally quite heavy and small in volume as it occupies little space, therefore, it is a high-density type of waste.

Biowaste comes from food leftovers that we throw away

This fraction includes small-sized plant remains left over from gardening and pruning (bouquets, grass). It does not include remains from tree pruning or similar (due to their greater size and woody nature) which is managed through the "pruning or green fraction" at the Recycling Points.

Organic waste has always existed throughout history, in small quantities, and it was either absorbed by nature itself or used as food by animals.

Most of this waste is generated in the home kitchen, *before* meals (peels, shells), *during* meals (inedible remains, bones, skins) and also, unfortunately, *after* meals (food that has expired or isin poor condition, or surplus food that ends up in the garbage can.) This type of food waste is also generated in commercial activities (greengrocers, grocery stores, markets, supermarkets, restaurant bars, hotels, among others), as well as in collective food centers (in schools, in companies). *See the next chapter about food waste.*

Within municipal waste, organic matter is the most unstable fraction as it is exposed to the action of microorganisms, which can degrade it biologically and then cause bad odors and leachates.

As we have seen, biowaste represents around 38%[156] of domestic garbage from homes and commercial waste. In other words: **almost 40% of the waste that we generate is organic matter.**

In 2012, in Catalonia,[157] a total of 384,136 tons of organic material was recycled (collected separately in brown bins). In Spain,[158] this figure was 547,564 tons and, in Europe, 28,540,000 tons. It is estimated[159] that in the same year the biowaste recycling rate from homes and businesses *was only* 22% of the total organic matter generated in Catalonia. The figures in Spain can't be contrasted.

---

156. My compilation from different studies: "Pesa la brossa" 2014. Study for the Polytechnic University of Catalonia and General Program of Prevention and Waste Management of Catalonia 2013-2020. According to the Agència de Residus de Catalunya 2014, the data are Organic 37%, paper, and cardboard 12%, glass 8%, plastics and metals 12%. "La gestió dels residus i el seu impacte en el canvi climàtic." Statistics 2014
157. Generalitat de Catalunya, statistics estadistiques.arc.cat
158. Annual report 2013 Ministerio de Agricultura, Alimentación y Medio Ambiente, Spanish Government
159. General prevention and waste management Program in Catalonia 2013-2020

Organic matter represents 40% of our garbage, but we only recycle 22%

There are 24 biowaste treatment facilities in Catalonia, while in Spain there are only 44. The difference is that selective waste collection (separately) of this fraction is mandatory in Catalonia. It should be noted that Catalonia is the only Spanish autonomous community where it is mandatory to collect organic waste separately, although it is widespread in the Basque Country.

The extension of the collection model, adding the fifth container for selective waste (organic matter or biowaste), contributes to increasing the global recycling rate by up to 40%. **Biowaste recycling is one of the keys in order to achieve a successful model** for waste management.[160]

The recycling of organic matter is one of the keys to success in recycling

---

160. Libro Verde *de la Sostenibilidad urbana y local en la era de la información.* Ministerio de Agricultura, Alimentación y Medio Ambiente, y Agencia de ecología Urbana de Barcelona, 2012

BENEFITS OF RECYCLING ORGANIC MATTER

The main benefits of recycling organic matter are:

- **Energy saving**: Recycling biowaste in facilities (anaerobic digestion) produces biogas, similar to that which is emitted in landfills, and enables the generation of energy

- **Saving resources**: Organic matter is converted into compost in the treatment facilities (composting process and anaerobic digestion). Compost is used as an organic fertilizer for agriculture and gardening and avoids the use of other fertilizers. Compost improves soil quality (fertility, porosity, water retention, and nutrient retention). Also, the fact that the other fractions – paper, glass, plastics, and metals – do not contain organic matter (which is easily degraded) it helps to improve the non-organic recycling, both in quantity and quality or efficiency

- **It improves the quality of air and water** by reducing pollution: Treating organic waste in recycling facilities avoids odor problems, as well as gas emissions and leachates in landfills and incinerators. Also, organic matter is one of the precursors of the generation of the aforementioned dioxins and furans in incinerators[161]

- **Improvement of soil quality**: The compost helps to improve the structure and fertility of degraded soils and deficiencies in organic matter[162] which are very common throughout the Spanish territory

---

161. "Biowaste management in the municipality," Ministerio de Agricultura, Alimentación y Medio Ambiente, Spanish Government
162. Ministerio de Agricultura, Alimentación y Medio Ambiente, Spanish Government

- **The emission of greenhouse gases decreases:** As we have said, landfill emissions contribute to the global warming of our planet. One of the great benefits of treating organic matter in recycling facilities is that it reduces the emission of gases such as methane CH_4 or CO_2 dioxide, responsible for global warming
- **Less waste is sent to landfill or for incineration**

The recycling of biowaste reduces the impact on the environment

Waste glass, paper, cardboard, plastic, and metals are recycled in order to take advantage of the materials and to minimize the environmental impact – less energy or water consumption, less air pollution or reduction of greenhouse gases. Unlike these materials, **organic matter is mainly recycled due to its high potential to have an impact on the environment**, despite it being a resource that is used as compost.

Want to know more?

▶ Biowaste composting facility (CAST)

https://www.youtube.com/watch?v=uLtrTkLHuBs

7.4.1. Food waste

Within the scope of waste prevention –generating less garbage –one of the current main battles is the huge quantity of food waste that we produce. We're *not aware* of it, but FAO[163] has established that, worldwide, 1/3 of the food produced for human consumption is wasted.

1/3 of the food that we produce is wasted

However, this waste does not only occur at the table and in kitchens, it affects the entire production chain, from the field to industry, transport, shops and homes. In terms of municipal waste, the level of waste produced by households equates[164] to 58% of the total, shops (groceries, markets, supermarkets, etc.) make up 26% and restoration (bars, restaurants, hotels, among others) the remaining 16%.

On a global scale, food production is 1.5 times greater than the demand for food, even though access to food is unequal.[165]

In Catalonia, almost 1.2 million tons of biowaste are generated each year, in the form of food waste, in homes and businesses in general. Each person wastes 6% of these foods, which end up in the garbage bin (selective or not) and this waste represents a total of 262,000 tons[166] per year. The average of this waste is 35 kg per inhabitant per year, or in other words: almost **100 grams per person and day.**

---
163. Food and agriculture organization of the United Nations
164. *Guide to avoid food waste,* Barcelona city hall
165. *Ibidem*
166. Agència de Residus de la Generalitat de Catalunya, residus.gencat.cat

The **economic** impact of food waste in Catalonia is estimated at €841M.[167] However, the impacts of waste can also be **social or ethical** (many people on this planet go hungry) and **environmental**: it's estimated that global food waste caused 3.3 Gt of CO_2eq., more than double the emissions produced by USA road traffic (cars, trucks, motorcycles, etc.) in 2010.[168] The global average emission per person per year due to food waste is equal to the emissions produced by a car during a 2,300 km journey (Barcelona-Copenhagen is 2,137 km). In other words: if food waste were a country, it would be the third greatest emitter of greenhouse gases.[169]

> If food waste were a country, it would be the third greatest emitter of greenhouse gases

The main problem[170] with food waste is that **people are not aware** of the act of throwing food away.

At the municipal level, there are many initiatives, such as Food Banks or Solidarity Points, where food surpluses are channeled to families who need it.

---

167. Área Metropolitana de Barcelona, www.amb.cat
168. Data found in FAO 2013 and "MalbaratamentAlimentari," Treball de recerca, AMB, 2014
169. http://www.residuosprofesional.com/desperdicio-alimentario-emisiones-gases/
170. "Malbaratament Alimentari," Treball de recerca, AMB, 2014

Want to know more?

▶ Great communication campaign for food waste in Ireland (ENG)

https://www.youtube.com/watch?v=VGTPKKOVoz4

▶ Good explanation of the food waste of Tristram Stuart in TED (ENG)

http://www.ted.com/talks/tristram_stuart_the_global_food_waste_scandal?language=es

▶ *Youtube* channel for food waste

https://www.youtube.com/playlist?list=PLWQMeO43vsucDxOZDQjGy746Oql_9PKv7

🛈 Official information about food waste. European Union (ENG)

https://ec.europa.eu/food/safety/food_waste_en

🛈 Communication campaign to become aware of food waste (CAT)

http://somgentdeprofit.cat/

🛈 *Guide to avoid food waste. City Council of Barcelona (CAT)*

http://ajuntament.barcelona.cat/ecologiaurbana/sites/default/files/Guia_per_evitar_malbaratament_alimentari.pdf

🛈 *Punt Solidari* project in El Prat del Llobregat (CAT)

http://www.elprat.cat/persones/serveis-socials/punt-solidari-servei-de-distribucio-gratuita-daliments

7.5. Benefits of waste recycling: A summary

We usually talk about the benefits for the environment brought about by us recycling waste, *but how much are we talking about?* Is recycling really good for the environment? Sometimes it's important to know how much we mean exactly, to put it in "black on white," with concrete cases and data.

Therefore, after reviewing each of the fractions, to facilitate a comparison of each of their benefits for the environment, here's the following table of **benefits for each fraction that we recycle:**

	GLASS	PAPER	PLASTIC	METAL	ORGANIC
Energy consumption	-20 to -30%	-70%	-84%	-75 to 95%	Saving
Raw material	1kg material x 1kg recycled	12 trees x 1 Tn	1kg oil x 1 kg	6 kg bauxite x 1 kg Al	Compost to improve soil quality
Resources needed in the industry	34% the needs	69% the needs	9% the needs		
Water pollution	-45%	-35% (-80% consumption)			
Air pollution	-20%	-74%	Emission reduction		
Global warming	-20% to 50%	Saving			Saving by landfill
Landfill and incineration reduction	Less waste to treat				

I have to confess that while I was drawing up the table, I was surprised to see the total reduction in resources and especially the consumption of energy, which we save by recycling our garbage. So much so, that **recycling is more a matter of saving energy and resources along with not polluting the planet.**

Recycling our garbage saves energy, resources and helps to conserve our planet

Conclusions

- Recycling waste saves over 30% of energy in the production of new packaging or materials

- Thanks to recycling, the consumption of raw materials is reduced (which are already scarce in Spain and Europe). The ratio is greater than 1 to 1. Recycling 1,000 kg of paper saves 12 trees from being felled

- Recycling reduces water consumption and pollution. Only 2% of the water on earth is potable

- Recycling contributes to reducing the emission of gases that cause global warming, *which is real and exists*

- And, above all, if we recycle waste, not only will we obtain direct benefits but also indirect ones, that is, we will avoid the inconveniences caused by landfills and incinerators

How important is recycling?

Want to know more?

▶ "Under The Dome" Documentary on China's Pollution, by Chai Jing. Viral phenomenon with more than 150 million visits (CH sub ENG)

https://www.youtube.com/watch?v=T6X2uwlQGQM

▶ Polluted air. *El Escarabajo verde* program RTVE (CAST)

http://www.rtve.es/television/20150429/soplo-aire-contaminado/1136279.shtml

8.
THE BEST WASTE IS NO WASTE

Not only must we recycle waste, but we must also reduce the waste that we generate since, from the environmental point of view, the best waste is that which is not generated at all. In the recycling chapter, we reviewed the waste strategy or hierarchy that we should follow (reduction, reuse, recycling, recovery, and disposal). Now, if recycling is necessary, waste reduction – or prevention – is even more significant (first option in the pyramid). We should even prioritize not generating or producing waste over recycling it. In a society where the culture of "use and throw away" is increasingly widespread and where products become waste in an instant, changing our habits is incredibly difficult.

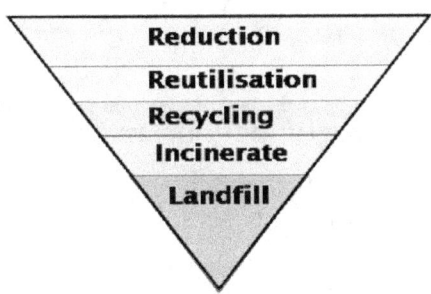

When I started to think about writing this book, I was in doubt whether to base it on waste reduction and prevention rather than recycling. However, how the *hell* would I explain that we have to reduce waste if we only recycle only 30-40% of our garbage? Who would read the book? So, I chose to address the basic principles of waste and explain the importance of recycling, and at least to dedicate a chapter (this one) to waste reduction.

The best waste is that which is not generated

Waste reduction, or prevention, is the set of measures or actions that are adopted before a substance, material or product becomes waste and that contribute to reducing: **the amount of waste** (including reuse and increasing the product lifetime), *its negative impact* on people or the environment and the content of *dangerous or harmful* substances.[171]

These measures or actions must be taken not only by the users or consumers of the goods but also by the industry and the manufacturers, whose responsibility in the different stages such as product conception, design, and distribution is essential.

---

171. *Estate waste prevention programme,* Ministerio de Agricultura, Alimentación y Medio Ambiente, Spanish Government

> **Prevention in the kitchen: The fried egg example**
>
> *This example that I've always liked*
>
> The fried egg is the product and the oil is the waste, these are the available options:
>
> **RECYCLE**: Oil waste is produced when cooking a fried egg in a pan. We can put this oil in a pot and take it to the recycling bin or Recycling Point.
>
> **REUSE**: Once the fried egg is cooked, we can save the surplus oil to reuse it, for example, to fry future eggs.
>
> **REDUCE**: Keep non-stick pans in good condition so that we hardly need any oil to cook the egg. If we're feeling particularly daring, we might even cook the egg in the oven or the microwave, thus reducing the use of oil (waste) as much as possible.

This example perfectly illustrates the options that exist in order to reduce the amount of waste that we generate; sometimes it's just a matter of **will**. Even though environmental awareness among consumers is increasing, other measures should be adopted so that companies and manufacturers incorporate specific waste prevention actions.

Something that stands out in the industrial field is the **planned obsolescence** of some products, which are designed to function for only a certain number of years thus forcing the consumer to buy again (turning them into a repeating customer). The sectors that should take urgent measures to extend the lifetime of their products are: those that manufacture household appliances (electrical and electronic equipment), furniture, textiles, tires, and packaging.[172]

The environmental benefits of waste prevention are greater than those of recycling because, if we don't produce that waste in the first place, we reduce energy consumption by 100%, natural resources are not used, water and air are not polluted, waste isn't sent to landfill or for incineration, among other benefits. However, we must also highlight the economic benefits that come from not needing to treat or recycle waste.

Between reducing and recycling there is also a middle ground: reusing waste or objects and, sometimes with a simple repair, an object can be reused again.

> The environmental benefits of reducing waste are greater than those of recycling

In the international arena, there is an exciting zero waste generation program. At the municipal level, there are several interesting initiatives, such as second-hand markets, Recycling Points for waste recovery, guides featuring shops for repair or resale, workshops to learn how to repair objects or even purchase-sale applications for used objects.

---

172. *Estate waste prevention programme*, Ministerio de Agricultura, Alimentación y Medio Ambiente, Spanish Government

Want to know more?

▶ *The incredible story of a plastic spoon, Greenpeace (ENG)*

https://www.youtube.com/watch?v=eg-E1FtjaxY

🛈 Bea Johnson's blog about how to live a zero waste life (ENG and FR)

http://www.zerowastehome.com

🛈 Lauren Singer's blog about how to live a zero waste life (ENG)

http://trashisfortossers.com/

https://d2pq0u4uni88oo.cloudfront.net/projects/916110/video-458662-h264_high.mp4

🛈 Catalan zero waste strategy, citizen initiative (CAT)

http://estrategiaresiduzero.cat/

▶ *Planned obsolescence,* ARTE, TV3, RTVE (ENG)

https://www.youtube.com/watch?v=24CM4g8V6w8

🛈 *Millor que nou 100% vell,* guide to repair shops, second-hand markets and workshops in the Metropolitan Area of Barcelona (CAST)

http://millorquenou.cat/es

▶ *Recuprat,* waste recuperation program at a Recycling Point, El Prat del Llobregat (CAT)

https://www.youtube.com/watch?v=P1TEvhR-FxY

9.
GLOBAL WARMING AND WASTE

The greenhouse effect is necessary for life on Earth. However, humans' actions are increasing this greenhouse effect, and the consequence is that the **Earth is warming up.**

Surrounding our planet, there is a layer of gases (mainly methane CH_4 and carbon dioxide CO_2) that works like the walls of a greenhouse: Sunlight shines on Earth, the gases allow ultraviolet solar radiation energy to pass through but block infrared or heat waves (when they reach the Earth, the ultraviolet rays rebound and transform into infrared rays).

The problem is that we have widened the layer of gases (*the thickness of the wall glass*), the greenhouse effect is increasing, less energy is escaping and the planet is heating up.

The greenhouse effect is necessary for there to be life on Earth

The six main greenhouse gases (GHGs) identified in the Kyoto Protocol that cause the greenhouse effect are:

- Carbon dioxide CO_2
- Methane CH_4
- Nitrous oxide N_2O
- Water vapor
- Ozono O_3
- Halocarbon gases HFC, PFC, SF_6, NF_3

The ensemble of these greenhouse gases (or GHG) is also known as "the carbon footprint," and corresponds to the total of the CO2 equivalence of each gas.

The global emissions[173] of GHG (greenhouse gases) in 2010 stood at almost 49 Gt[174] of CO_2 eq., practically twice as much as in the year 1970, when there were 27,000 tons.

The total GHG emissions[175] in Catalonia, in 2012, exceeded 43 Mt of CO_2 eq. In Spain, it was more than 340 and, in Europe (EU-28), 4,544 Mt[176] CO_2 eq. Each Catalan emitted an average of almost 6 tons of CO_2 eq. in the year 2012.

The emission of greenhouse gases has doubled since 1970

The sector that contributes the most to the greenhouse effect is energy (energy processing), and it can easily be called "The Sector," since it contributed over 75% of the total GHG[177] in Catalonia in 2012. It is followed by the agricultural sector with almost 10%; industry, with over 8%; and, in fourth place, emissions related to waste (treatment and management), with 6% in 2012, which in 2014 fell[178] to 3%, and represents almost 0.8 Mt CO_2 eq. GHG emissions from waste are mainly methane CH_4 (92%), and the majority (89%) are produced in landfills.[179] The waste that we generate contributes to the warming of the Earth, through the greenhouse effect.

---

173. *Climate Change 2014, Mitigation of Climate Change, Summary for Policymakers and technical summary,* INTERGOVERNMENTAL PANEL ON climate change (IPCC). UNEP. ONU
174. 1 Gt or Gigaton = 1,000,000,000 tons = 1,000,000,000,000,000 kg = 10^{15} kg
175. "Fifth report on the Catatonia progress to Kyoto objectives," Oficina Catalana del Canvi Climàtic
176. 1 Mt or Megaton = 1,000,000 tons = 1,000,000,000 Kg = 10^9 kg
177. "Fifth report on the Catatonia progress to Kyoto objectives," Oficina Catalana del Canvi Climàtic
178. "Waste management and his impact on Climate Change," Statistics 2014, Agencia de Residus de Catalunya
179. "Fifth report on the Catatonia progress to Kyoto objectives," Oficina Catalana del Canvi Climàtic

Stop garbage

The waste that we generate contributes to the warming of the Earth

Thanks to the correct management of some of our waste, it was possible to reduce these emissions, avoiding[180] almost 0.7 Mt CO2 eq. in 2014. However, in 2010, the emissions from waste treatment and management were almost double those emitted in 1990 (93% increase).

Below, you can see the specific amounts (in GHG) that we're talking about when considering the benefits provided by recycling for each waste fraction:

	AVOIDED GHG* (1kg CO_2 per 1 kg of)	Equivalences** (Km by car per 1 kg)
Glass	0.31	2
Organic	0.80	5
Plastic	1,.14	7
Steel	1.62	10
Paper	2.94	17
Aluminium	8.87	52

**GHG AVOIDED[181] *Equivalence[182]

For each kilogram of the different fractions, the table shows the equivalence in kilometers that a car can travel

---

180. "Waste management and his impact on Climate Change," Statistics 2014, Agencia de Residus de Catalunya
181. "Municipal Solid Waste Generation, Recycling, and Disposal in the United States," Facts and Figures for 2012, United States Environmental Protection Agency
182. Factor determined on: Savings of CO2 emissions in the products of the purchase net reciclat. Car type EURO 4

by producing the same emission of GHG. For example, **recycling 1kg of paper would be equivalent to avoiding car emissions for 17 km** (from Barcelona to El Prat Airport). What's remarkable is that aluminum recycling is the type that prevents the most GHG gases.

In conclusion, separating waste into different fractions or waste bins for its subsequent recycling and composting is the best option —in terms of waste management and treatment —in order to reduce the greenhouse effect of the planet,[183] with the lowest net GHG flow.

Waste separation and recycling is the best option in order to reduce the global warming of the Earth

---

183. "Waste management options and climate change," European Comission

Want to know more?

▶ *An inconvenient truth,* Oscar-winning film by the Hollywood Academy, of the ex-US Al Gore *(What a great president he would have been!)* Full movie (ENG and CAST)

http://www.mojvideo.com/video-an-inconvenient-truth-2006-full-movie-vostfr-slo-podn/32f7a8034ed710edce13

https://vimeo.com/36522029

▶ Movie trailer (ENG)

https://www.youtube.com/watch?v=Bu6SE5TYrCM

▶ Leonardo Di Caprio at the 2014 Climate Summit at the UN (ENG)

https://www.youtube.com/watch?v=vTyLSr_VCcg

▶ Good explanation of global warming, National Geographic (ENG)

https://video.nationalgeographic.com/video/101-videos/climate-101-causes-and-effects

i Climate change (ENG):

http://www.un.org/climatechange

https://www.nationalgeographic.com/environment/global-warming/global-warming-overview/

10.
ECONOMY AND WASTE

Waste recycling not only has positive effects on the environment but also at an economic level, since it contributes to the GDP, and at a social level, as a generator of employment.

10.1. GDP and waste

Waste, just like any activity in our society, is a part of the economy. It contributes to the generation of wealth, produces goods and services, and also generate employment.

What is GDP?

GDP (Gross Domestic Product) is a concept that is used in macroeconomics to measure the operations and flows of the economy in a country or region, in order to obtain an overall vision.

In more technical words, GDP is the monetary market value of the production of physical goods and final services carried out over a year's economy by the productive factors in a country.

How much does the waste sector contribute to the GDP?

In Catalonia, it is estimated that its contribution for the year 2013 was around €12,000 M, that is, 6% of the Catalan[184] GDP. In Spain, it is estimated at almost €11,000 M for the year 2011, which is 1% of the Spanish[185] GDP. In Europe[186](EU27), the contribution of the waste sector is estimated to be over €130,000 M/year for the year 2012, which represents 1% of the European GDP. According to the United Nations Environment Program (UNEP), the global waste market[187] is estimated to be at €410,000 M for the year 2011, which would represent 0.6% of the world GDP.

---

184. Generalitat de Catalunya, gencat.cat
185. Only the data of 3.6% is available for the entire environmental sector. It is assumed that 30% of contribution is due to the waste sector
186. 1% on "Resources and waste," 2012, and 0.75% found in "El Medio Ambiente y Europa" SOER 2011, both studies from European Environment Agency
187. General prevention and waste management Program in Catalonia 2013-2020

The waste sector accounts for 6% of the Catalan GDP, and 1% of the Spanish and European GDP

Different companies participate in the management and treatment of waste to generate this economic activity in the waste sector: in Catalonia,[188](2013), 900 companies; in Spain,[189](2010), a total of 2,400 companies. In Catalonia,[190] in addition to these 900 waste treatment and direct management companies are the 2,900 waste transport companies, as well as those that manufacture machinery and those involved in engineering and consulting services.

Relationship between GDP and waste

Historical data indicates that in most industrialized countries there is a direct relationship between waste generation and economic activity.

Over time, and specifically as a result of the progress experienced in the twentieth century from both the technological point of view and the development of society and the economy, the GDP increased 23 times, mineral extraction 27 times, fossil fuel consumption 12 times and the total extraction of materials 8 times, among other data.[191]

---
188. Generalitat de Catalunya, gencat.cat
189. "Economic study about the environmental sector in Spain 2011," Fundación Fórum Ambiental.
190. General prevention and waste management Program in Catalonia 2013-2020
191. "Decouple the use of natural resources and the environmental impacts of economic growth," 2011, ONU-UNEP

Equally, if we examine the GDP of different European countries and the generation of municipal waste (a rough indicator of partial consumption of resources) we see:

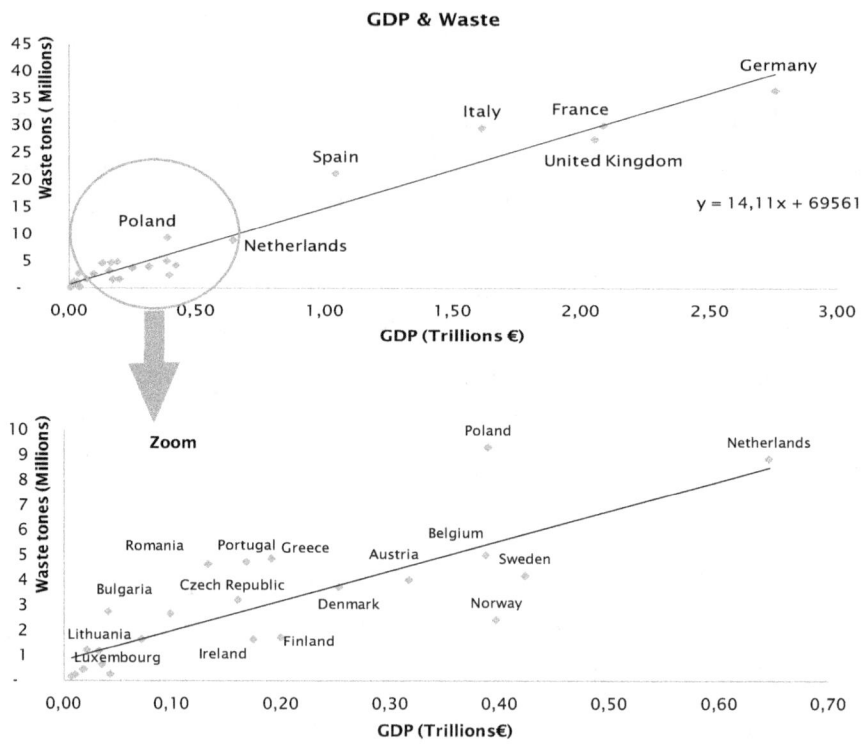

Source: EUROSTAT official data, GDP and municipal waste for 2012

This graph shows a clear general trend: the higher the level of wealth or development of a country (GDP), the greater its waste generation.

This trend implies that **if a country wants to progress in its development, it will generate more waste**, with the current model. For example, if Spain (with a GDP = 1 trillion Euros and generating 21 million tons/year) proposed to increase its GDP by an additional 1 trillion, equaling the United Kingdom, the municipal waste generation would increase by more than 6 million tons/year (up to 27 million tons/year).

> There is a direct relationship between the increase in wealth or GDP of a country and the increase in waste

Due to the shortage of the world's natural resources, which as I said before are finite, the overexploitation of materials is one of the main priorities worldwide, both for the UN and for the EU.[192] Therefore, it is necessary to develop economic activity and production (to achieve human well-being) by reducing both global consumption and the impacts of the extraction of natural resources.

All of these reasons make it important to stand firm and work towards waste reduction (*see chapter 8*), the dematerialization and efficiency of the economy, and the development of new concepts or the use of tools such as the circular economy (*see chapter 10.4*).

---

192. "Europe 2020," the EU's growth strategy for the next decade

10.2. Recycling is cheaper than not recycling

Municipalities, on the one hand, receive income from recycled waste, while on the other, they are paying taxes for waste that is not adequately segregated, the rejection fraction that ends up in incinerators or landfills.

The waste that is deposited in the rejection fraction container, which is usually gray, not only does not provide any income but actually causes additional costs. As in many European countries, in Catalonia, tax has to be paid on waste that goes for incineration and to landfill: €9/ton for incineration and €19/ton in the case of landfill (2015).

Also, it's estimated that waste and materials sent annually to European landfills could have a total commercial value of around[193] €5,250 M.

Tax has to be paid on non-recycled waste from the gray container that ends up in incinerators and landfills

---

193. "Being wise with waste: the EU's approach to waste management," European Comission

One of the main revenues related to municipal waste comes from the sale of paper and cardboard from the blue container. In most municipalities, there are agreements with guilds of recovery associations that buy this material in exchange for financial compensation. It should be noted that the process is very transparent, well-organized and traceable.

> The more the recycling, the more the revenue generated for town councils and, at the same time, the lower the expenses

Collection and Recovery System or CRS (non-profit entities, although managed by large production companies) also generate income for the municipalities. According to the law, waste producers have to take responsibility for the waste generated by their commercial activity, in line with the **polluter pays** principle, which in the waste sector is also called the *extended responsibility of the producer*. In an orderly, transparent and efficient manner, these producers come to an agreement with the municipalities to take care of waste collection in exchange for economic income. In other words, they transfer their responsibility in exchange for payment for the services rendered.

In any case, don't forget that these revenues are generally provided by the customers or citizens who buy these products, as companies have included the tax incurred for each piece of packaging that they put on the market in the final price. Among these CRS, the following are worth highlighting: Ecoembes (for plastic, metal, and mixed containers), Ecovidrio (glass containers), Ecopilas (batteries and accumulators), Ambilamp (bulbs and fluorescents), OFIRAEE (electrical and electronic equipment from the Clean Points), SIGRE (medicines) and Signus (tires). The income that these companies contribute to the municipalities varies, in the sense that **the more the recycling, the more the income.** I'd like to add that many of these revenues —controlled and supplied by the CRS— do not cover the costs involved in the collection of waste.

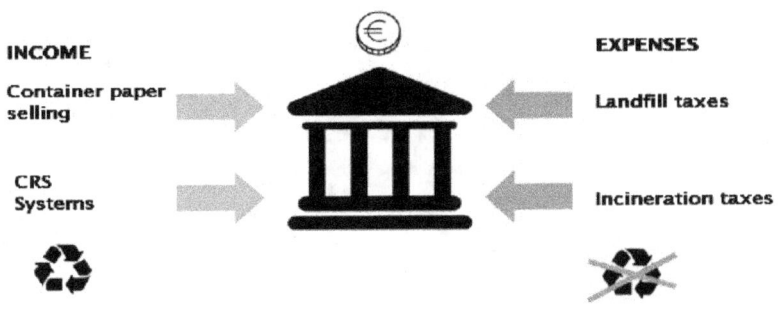

In summary, one could say that the waste that goes into the recycling containers does not imply a direct cost to the citizen and also generates income for the municipalities which is proportional to the tons of waste recycled. In contrast, non-recycled waste that ends up in landfills or incinerators not only does not produce an income but its treatment is taxed.

The cost of mixed waste treatment is more expensive than the cost of recycled waste treatment

The treatment of waste from the rejected or mixed fraction costs about €89/ton,[194] while if this is deposited into the brown container to make compost, it only costs €51/ton. The treatment of waste from paper, packaging and glass containers implies no cost to the city councils –or to the citizens – since the companies pay for it through the mentioned CRS.

---

194. Public price for the urban waste treatment service for elimination in the Provincial Council of Bizkaia, Udaltalde City Council Study

Economy and waste

Want to know more?

🔖 ▶ Campaign of the Udaltalde City Council: "Why throw money away?"

http://ut21.org/noticias_mas-c.php?id=609

🔖 CRS websites:

 https://www.ecoembes.com

http://www.ecovidrio.es/

http://www.ofiraee.es/

http://www.sigre.es/

http://www.signus.es/

http://www.ecopilas.es/

http://www.ambilamp.es/

10.3. Recycling creates employment: less garbage, more employment

The environmental recycling benefits that we have seen in previous chapters may fail to convince those who lack a developed ecological understanding or awareness of the environment. However, everyone is aware of the importance of employment, especially after witnessing the last economic crisis.

In this sense, it's a fact that the waste sector currently creates jobs and direct employment: in Catalonia,[195] around 28,000 people in 2013 (0.7% of the active population); in Spain,[196] over 140,000 people in 2010 (representing 0.6% of the active population); and in Europe[197] (EU27), a total of 1.8 million jobs according to 2006 data; in the USA this figure was 1 million in 2002. Of course, data regarding indirect employment must also be added here.

---

195. Generalitat de Catalunya gencat.cat
196. "Economic study about the environmental sector in Spain 2011," Fundación Fórum Ambiental
197. "More jobs, less waste," Friends of Earth, 2010

The treatment of recycled waste involves different types of operations that are more complex than simply burying or burning garbage and therefore requires more work positions. The following table shows the jobs created in the USA[198] and in the European Union[199](EU) for every 10,000 tons of waste treated per year, depending on the operations carried out:

	Jobs per 10,000 tons/year	
	USA	EU
Waste recycling		250
Material recovery facility	10	
Paper recycling facility	18	
Glass recycling facility	26	
Plastic recycling facility	93	
Compost facility	4	
Landfill	1	20-40
Incineration	1	10

Although the data is disparate, it's clear that in the USA the scale of jobs, in a material recovery plant is x10 times higher than in landfill or incineration and this increases significantly in the recycling plants (paper x18 times, glass x26 or plastic x93). The European data follows the same increasing employment scale, x10 times.

---

198. "More jobs, less waste," Friends of Earth, 2010, and also found in "The waste incineration in figures," Greenpeace, 2010
199. "A step forward in the sustainable consumption of resources: thematic strategy on the prevention and recycling of waste, "Communication from the European Commission to the European Parliament, Brussels 2005

Waste recycling increases employment generation x10 times

What's more, employment growth in the Metropolitan Area of Barcelona[200] has been verified. In the past, when most of the waste was taken to landfill (in Sant Joan Valley, Garraf), around 80 people were working, nowadays there are almost 800 people working in the treatment and waste management plants. Once again, we find the same multiplier factor, x10 times, in employment generation.

---

200. "Waste generates employment," *El Periódico de Cataluña*," 17 January 2015

As we have already mentioned, 42% of the waste generated in Europe is recycled, and 1.8 million people are employed in the recycling sector. Just think, if a recycling level of 70% was attained, over 500,000 new jobs[201] would be created (remember some of the unemployment figures for 2016: Catalonia[202]= 659,600 people, Spain[203]= 4,850,800 people, and Europe[204]= 22 Million people).

Recycling 70% of EU waste would create 500,000 jobs

The recycling of the organic[205] fraction presents a similar opportunity. Remember that, at the moment, in Spain, only 17% of organic matter is collected separately and it is estimated that this generates almost 11,500 direct jobs. Well, if 80% of the organic matter generated in households and from commercial activities was selectively collected and treated, almost 5,200 jobs would be created, so we would achieve a total of 16,700 jobs related to the organic recycling fraction.

---

201. "More jobs, less waste," Friends of Earth, 2010
202. Data INE 3th trimester 2015
203. *Ibidem*
204. EUROSTAT, November 2015
205. "The employment generation in biowaste management within the framework of the generalization of the selective collection," Trade Union Institute of Labor, Environment, and Health (ISTAS) of Workers' Commissions (CCOO)

Want to know more?

1 Summary of Friends of the Earth report, "More work, less garbage" (ENG)

http://www.foeeurope.org/press/2010/Sep14_half_million_new_jobs_could_be_created_by_recycling_more.html

1 "The generation of employment in the management of the organic matter of urban waste within the framework of the generalization of the selective collection," Trade Union Institute of Labor, Environment, and Health (ISTAS) of Workers' Commissions (CAST)

http://www.istas.net/web/abretexto.asp?idtexto=4071

1 "The waste incineration in figures," Greenpeace, 2010 (CAST)

http://www.greenpeace.org/espana/Global/espana/report/contaminacion/100720.pdfxx

10.4. Waste as resources: moving towards a circular economy

The availability of resources that we extract from the planet is decreasing, and by contrast, the consumption of resources is increasing, especially in developed countries. Therefore, the availability of (non-renewable) resources is one of the crucial challenges facing our society in order to, on the one hand, be able to guarantee the quality of life and survival of its inhabitants, and on the other hand, to have competitive and sustainable economies, in the not too distant future. We need resources.

The production of products or goods has thus far occurred linearly where: raw materials are extracted; products are made; consumed, and finally, they are rejected and generally sent to landfills. For some years now, a concept that aims to turn linear production into circular production has been on the scene. This new idea sees waste become a resource which can be used to produce new goods and services, and thus close "the product's lifecycle." We are imitating the biological cycle of nature. This new concept is called "circular economy."

Stop garbage

LINEAR ECONOMY

EXTRACTION  PRODUCTION  THROW

CIRCULAR ECONOMY

EXTRACTION   PRODUCTION   Ecodesign
Efficiency
Increasing life
Business model

**REUTILIZATION
REPAIR
RECYCLING**

The products of today can become the resources of tomorrow

Some of the principles upon which circular economy is based are the following:

• Global economic model: The aim is to untie economic growth and the consumption of finite resources, with the aim of developing a resilient economy that works in the long term.

• Eco-conception: The design is aimed at the efficient use of materials, considering the environmental impact of products during their life cycle, integrating this into their conception (to be able to recycle them or take better advantage of them). The products have a long life, can be disassembled, are easily repairable, etc.

• Economy of opportunities and functionality: This involves the promotion of new opportunities regarding design, products and services and business models. For example, turning a product into a service (such as the rental of blue jeans http://www.mudjeans.eu/). Another example are the eco-industrial parks, where some industries have and produce waste that could be raw material for others.

Of course, we must also add the principles that we have already discussed such as reuse, repair or recycling.

Waste recycling helps to reduce the pressure on the environment that creates the need for resources and raw materials in order to manufacture products. In 2006,[206] waste recycling covered 41% of paper consumption, 42% of iron and steel consumption, 14% of glass consumption and 4% of plastics consumption.

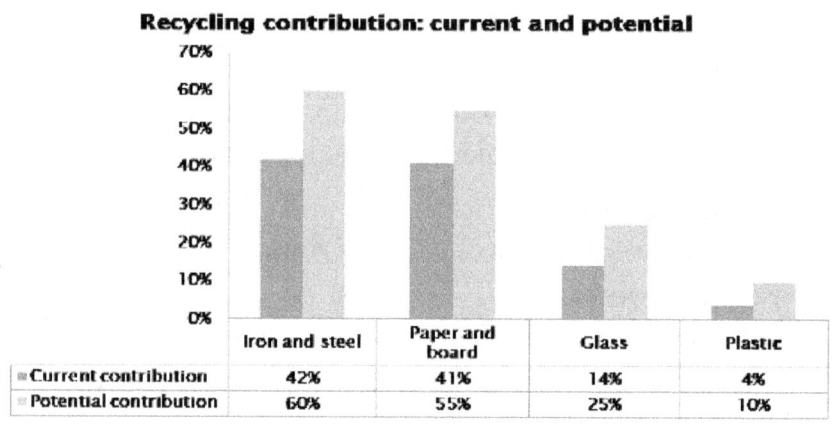

By recycling more, we could increase the requirement coverage for some resources, but not for all of them. The potential for consumption coverage for iron-steel is 60%, for paper it's 55%, for glass 25% and plastics around 10%.

Recycling helps to reduce the pressure on our environment, but it is not enough

---

206. "Income, employment, and innovation: the role of recycling in a green economy," Study of the European Environment Agency, Copenhagen, 2011, published by the MAGRANA Ministry

A conceptual change is needed, from the current linear economy to a circular economy. It certainly won't be easy and it requires everybody's collaboration: civil society, politicians and the business world. Promoting sustainable production and consumption policies, reorienting production processes and developing new business models is vital.

The circular economy benefits the environment (stops climate change, saves energy and water, reduces the emission of polluting products) and society (generates employment, fosters opportunities for new businesses, encourages growth and competitiveness).

Want to know more?

▶ European Parliament video about circular economy (ENG):

http://europarltv.europa.eu/es/player.aspx?pid=b14e4401-dea5-4b47-ac67-a517009f495e

▶ Ellen MacArthur Foundation video (ENG):

https://www.youtube.com/watch?v=zCRKvDyyHmI

▶ "Circular economy" *El Escarabajo verde* TV documentary RTVE (CAST)

http://www.rtve.es/alacarta/videos/el-escarabajo-verde/escarabajo-verde-economia-circular/2828228/

🛈 Information (ENG)

http://ec.europa.eu/environment/circular-economy/index_en.htm

http://www.ellenmacarthurfoundation.org/

EPILOGUE: RECYCLING IN BRIEF

To focus the problem of garbage, we must bear in mind that waste is a problem of **scale** and **order**. *Order*, in the sense that we all like our house to be in order, for everything to be in its place and for there to be a certain level of organization. In terms of *scale*, I mean that every year the number of people on the planet increases and we've reached the point where there are so many of us and we are consuming so many of our planet's finite and limited resources, that we are beginning to seriously alter our environment and cause irreversible changes. Scientists have announced that our planet Earth has already entered a new geological era, called *Anthropocene*.[207]

---

[207]. "Human impact has pushed Earth into the Anthropocene, scientists say," *The Guardian* 7 January 2016 https://www.theguardian.com/environment/2016/jan/07/human-impact-has-pushed-earth-into-the-anthropocene-scientists-say

My second reflection is related to why it's important to recycle. In a nutshell, I would say **to avoid the negative impacts on the environment** due to global warming and, of course, the pollution of the atmosphere and water, and similarly, to avoid the growing **scarcity of resources** on our planet, which can be solved, to some extent, by recycling the garbage that we generate.

> Recycling cares for our environment, guarantees quality of life for us and future generations, and... it makes us feel happier

I hope that you liked this book, that you enjoyed yourself and that it has awakened in you, at least, some curiosity about *the exciting world of recycling*. To learn more about recycling, garbage or waste, and to keep yourself up-to-date regarding these issues, I hope to see you on the blog www.stopgarbage.com

Thank you for the time that you have dedicated to *Stop Garbage*. If you liked it and think it has been useful in helping you to understand the recycling world, I would appreciate it if you could leave your opinion on Amazon. Your support is very important so that I can continue to explain the importance of recycling our waste. I promise to read all of your opinions, and I will try to give *feedback* to improve the book. You can leave your opinion on the book's Amazon page in the section "customer reviews" www.amazon.com.

Thank you for your interest, time and attention.

Best Regards.

Alex

Acknowledgments

To Ignacio García-Bermúdez, for his wonderful cover illustration. If you liked it, you can see more of his work at:

https://www.instagram.com/igb78/

To Maria Pfaff for her great ideas and contributions on the back cover

To all my teachers and trainers, my parents and my brothers, my bosses in different jobs, especially Antonio Boscadas, who has always believed in me and from whom I continue to learn

Jordi Figueras, municipal technician of the Barcelona City Council, who taught me the passion and knowledge of the waste and street cleaning sector

To the City Council and the citizens of Prat del Llobregat, I thank them for the work they provide me with and their great commitment to waste generation that has led me to improve, understand and explain the management of municipal waste

To my colleagues, to the municipal technicians who give me many experiences of improvement and who do a great job in the shadows

To my friends who are always there

This book has been printed on-demand, only ordered units are produced, wich reduces excess production.

The used ink is chlorine-free and the acid-free interior paper is FSC certified (see page 74).

We hope that Amazon will soon facilitate 100% recycled white paper book option.

www.ingramcontent.com/pod-product-compliance
Lightning Source LLC
Chambersburg PA
CBHW051314220526
45468CB00004B/1340